Composite Fouling on Heat Exchanger Surfaces

COMPOSITE FOULING ON HEAT EXCHANGER SURFACES

HONG YU

Nova Science Books
New York

Library of Congress Cataloging-in-Publication Data:
Hong, Yu.
Composite fouling on heat exchanger surface / Hong Yu.
p. cm.
Includes index.
ISBN 1-60021-116-X
1. Heat exchangers--Fouling. 2. Sugar machinery--Protection. 3. Calcium silicates. I. Title.
TJ263.H66 2007
621.402'5--dc22 2006008453

Published by Nova Science Publishers, Inc. ✣ New York

Contents

ABSTRACT

Deposition of amorphous silica (SiO_2) and calcium oxalate (CaOx) on the calandria tubes of juice evaporators cause serious processing problems in Australian cane sugar mills. The removal of these deposits by mechanical and chemical means is a time-consuming and costly experience. The cost of downtime and chemical cleaning can be several million dollars per year for the Australian sugar industry. The interactions between CaOx and SiO_2 have not been investigated previously because conventional studies only address fouling by individual components. The present work evaluates their interactions using two experimental approaches: batch tests for assessing kinetic and thermodynamic behaviour, and fouling-loop experiments for examining composite fouling behaviour under different operating conditions. The above two approaches were employed both in the absence and in the presence of sugar to elucidate the effect of sugar on composite fouling mechanisms and to determine the controlling species responsible for composite fouling. The composite fouling experiments were performed in a novel closed-loop circulation system simulating the effect of feed composition of successive stages of evaporation cycle in a single run. In addition, the fouling-loop system was operated in a constant composition mode to study the effects of thermal hydraulic conditions on composite fouling. The combined information obtained from both the batch and fouling-loop tests in this study offer a unique insight into the mechanisms of composite fouling of CaOx and SiO_2. Some of the highlights of the obtained results are as follows:

- Identification of a complex interactive process in calcium oxalate monohydrate – silica (COM-SiO_2) systems by investigation of the kinetics and thermodynamics of COM-SiO_2 coprecipitation in water and sugar solutions, and an understanding of the mechanisms of these interactions;
- Development of a novel fouling-loop system, which is simple, efficient and cost effective for the study of the effect of juice composition on scale formation in various stages of juice evaporation;
- Elucidation of composite fouling mechanisms, e.g., a feed composition dependent fouling mechanism is proposed;
- Isolation and verification of the existence of certain species in composite deposits, which is known to be thermodynamically unstable. In other words, it is established that calcium oxalate trihydrate is stable under certain conditions;

- Evaluation of the role of thermal hydraulic operating parameters in determining the characteristics of subcooled flow boiling heat transfer and in determining the strength of the composite deposit;
- Development and validation of an empirical model to predict the subcooled flow boiling heat transfer coefficients in water and sugar solutions;
- Development of an analytical model incorporating the effects of operating parameters for COM and SiO_2 composite fouling in sugar solutions. This model predicted the experimental data better than available models.

Results of this work are significant, not only because they have made a valuable contribution to advance the fundamental understanding of heat exchanger fouling, but also because they may play a key role in the development of scale control and removal strategies to minimize the composite fouling in Australian sugar mill evaporators. For example it was found that, in order to effectively minimize the rate of composite fouling and reduce the scale tenacity, it would be necessary to control thermal hydraulic operating conditions, especially the fluid velocity, and to adjust the initial CaOx/SiO_2 supersaturation ratio to the optimum value. To achieve the optimal CaOx/SiO_2 ratio, certain device can be developed to sequentially measure oxalic acid and SiO_2 concentrations in juice so that the correct proportions of chemicals can be added. Model simulations of the composite fouling rate may also effectively and economically provide comparative and relevant information essential for process optimisation and evaporator design.

PREFACE

The purpose of this book is to present a summary of some of the latest developments on the fundamental understanding of the complex physico-chemical phenomenon of *composite fouling* present on the heat exchange surface of sugar mill evaporators.

Deposition of amorphous silica (SiO_2) and calcium oxalate (CaOx) on the calandria tubes of juice evaporators cause serious processing problems in Australian cane sugar mills. The removal of these deposits by mechanical and chemical means is a time-consuming and costly experience. The cost of downtime and chemical cleaning can be several million dollars per year for the Australian sugar industry. The interactions between CaOx and SiO_2 have not been investigated previously because conventional studies only address fouling by individual components.

Based on a PhD research program carried out by the author, this book presents a unique approach to investigating the mechanisms of CaOx-SiO_2 composite fouling using two experimental approaches. First, the thermodynamics and kinetics of CaOx and SiO_2 co-precipitation in water and sugar media were evaluated to elucidate the mechanisms of co-precipitation and to identify the controlling species for co-precipitation kinetics. Second, the composite fouling experiments were performed in a novel dynamic set-up simulating the effect of feed concentration of successive stages of evaporator system within one run. In addition, the dynamic system could operate in a constant composition mode to study the effects of thermal hydraulic operating conditions, including sucrose concentration, temperature and velocity, on composite fouling behaviour. The combined information obtained from batch and dynamic tests then form the basis for recommending optimal design and pre-treatment strategy, as well as a predictive model for the composite fouling of COM and SiO_2.

Principal audiences of this book are likely to include industrial and academic researchers as well as postgraduate students in heat transfer, heat exchange fouling, crystallization, crystal growth, and particle technology areas. The book also aims to provide a useful source of reference for practicing professionals such as process engineers and chemists in sugar industry. Therefore, this book should serve the dual purpose of being used for both research and professional reference. Relevant subjects of this book include composite fouling, heat exchange surface fouling, boiling heat transfer, crystallization, mix-salt co-precipitation, silica chemistry, separation and colloid and interface science, surface adhesion, sugar and particle technology.

Results of this work are expected to make a valuable contribution to advance the fundamental understanding of heat exchanger fouling, as well as in the development of scale control and removal strategies to minimize the composite fouling in Australian sugar mill evaporators. For instance, it was found that, in order to effectively minimize the rate of composite fouling and reduce the scale tenacity, it would be necessary to control thermal hydraulic operating conditions, especially the fluid velocity, and to adjust the initial $CaOx/SiO_2$ supersaturation ratio to the optimum value. To achieve the optimal $CaOx/SiO_2$ ratio, certain device can be developed to sequentially measure oxalic acid and SiO_2 concentrations in juice so that the correct proportions of chemicals can be added. Model simulations of the composite fouling rate may also effectively and economically provide comparative and relevant information essential for process optimisation and evaporator design.

The author would like to express his gratitude to Dr. Roya Sheikholeslami of AAA Water Energy Technologies Pty Ltd for her advice and guidance throughout the work. Working with Dr. Sheikholeslami has been a very fruitful learning and research experience. The author wishes to extend his gratitude to Dr. Bill Doherty of Sugar Research Institute, Australia for his valuable support and suggestions, as well as his many visits to Sydney throughout the program. Without the helps of Dr. Sheikholeslami and Dr. Doherty, this book would not have been written.

The author is sincerely grateful to Dr. Peter Wright, Dr. Peter Worland, and Professor Paul Watkinson for examining the manuscript and many enlightening comments.

The author would like to thank Australian Research Council, University of New South Wales and Sugar Research Institute for their financial support of the program. The author is also indebted to all who supported and have been involved in various stages of this project for their contributions.

Hong Yu
December, 2005

Chapter 1

INTRODUCTION

1.1 CALCIUM OXALATE AND AMORPHOUS SILICA

1.1.1 Calcium Oxalate

Calcium oxalate (CaOx) occurs widely in nature in many plants and animal organs as an important product of biomineralization (Arnott, 1982; Franceschi and Horney, 1980). Precipitation of calcium oxalate is a major cause of kidney stones and fouling in pipes and processing equipments in pulp and paper and sugar industries (Nancollas, 1982; Ulmgren and Radestrom, 1999a).

Calcium oxalate is sparingly soluble in water and its solubility increases with increasing temperature and ionic strength, but is little affected by pH at pH > 4 (Stephen and Stephen, 1963; Streit et al., 1998; Ulmgren and Radestrom, 1999a). Calcium oxalate can also be rendered soluble in the presence of calcium-sequestering agents such as citrate and ethylenediamine-tetraacetic acid (EDTA) (Franceschi and Horney, 1980; Ryall, 1994). The crystal growth of calcium oxalate from aqueous solution is shown to follow a second order kinetics with the activation energy in the range of 25-50 kJ/mol (Garside et al., 1982). The process of crystal growth of calcium oxalate can often be inhibited by the presence of various additives in solution such as polyelectrolyes and surfactants (Sikiric et al., 1999; Sutor, 1969). These inhibitors appear to act in three ways: by complexing calcium ion as mentioned earlier, by forming macromolecular colloids with calcium oxalate to stabilise the crystals in solution (steric stabilization, see section 2.3.2.2), or by adsorption on the faces of growing crystals thereby blocking the nucleation and growth sites (Crawford et al., 1968; Millan et al., 1998).

Calcium oxalate is known to form three hydrates in aqueous solution – calcium oxalate monohydrate ($CaC_2O_4{\cdot}H_2O$, COM), calcium oxalate dihydrate ($CaC_2O_4{\cdot}2{+}\chi H_2O$ ($\chi<0.5$), COD) and less commonly, calcium oxalate trihydrate ($CaC_2O_4{\cdot}3H_2O$, COT) (Nancollas, 1982; Sheehan, 1981). Each of these phases has different physico-chemical properties, including solubility, stability, kinetic behaviour and conditions of crystal formation.

The solubilities of the three hydrates in aqueous solution are found to decrease in a sequence of COT > COD > COM (see Table 1.1). The least soluble COM is the only thermodynamically stable phase at high temperatures ($\geq 60^{o}C$) and is generally obtained with similar concentrations of calcium and oxalate ions in solution. The formation of the more soluble COD and COT, on the other hand, is favoured by high initial $[Ca^{2+}]/[C_2O_4^{2-}]$ ratios, stirring dynamics, low temperatures and the presence of other species such as carboxylic acids (Callejas-Fernandez et al., 1993; Cody and Cody, 1994; Skrtic et al., 1987). Both COD and COT are kinetically favoured in the initial stage of crystallization due to the Ostwald-Lussac rule but readily undergo phase transformation into COM via a dissolution – recrystallization process (Skrtic et al., 1987). The kinetics of hydrate transformation is highly sensitive to temperature and the ionic medium composition (Nancollas, 1982) (see section 4.3 for more details).

Table 1.1 Solubilities of calcium oxalate hydrates in water at 25°C (Streit et al., 1998)

Calcium oxalate hydrate	Solubility (ppm)
COM	6.0
COD	11.1
COT	13.8

1.1.2 Amorphous Silica

Silica, the most abundant compound found in the earth crust, is an important media for global material cycle. It occurs in rocks, soils, organisms and natural water systems (ground water, seawater, etc.) in various forms such as silica minerals, silicate minerals and amorphous silica (Krumbein and Werner, 1983). It is also a major contributor to scaling in process industries such as pharmaceutical, geothermal and desalination industries where silica-bearing waters are concentrated or used (Chan, 1988; Henley, 1992)

An important aspect of silica chemistry concerns the silica-water system. When dissolved in water, silica is weakly ionised and is generally considered to exist in the form of undissociated silicic acid (H_4SiO_4, reactive silica) in most naturally occurring waters up to pH ~ 8.0 according to the following ionic equilibrium:

$$SiO_2 + 2H_2O \Leftrightarrow H_4SiO_4 \qquad (1.1)$$

Table 1.2 gives the solubility of amorphous silica in water at different temperatures. The variation of silica solubility at each temperature arises from differences in the particle size, degree of hydration and level of trace impurities, especially of divalent and trivalent cations such as magnesium, aluminium and iron (Iler, 1979; Patterson, 1993). The effect of trace metal is believed to be due to the formation of an insoluble metal silicate layer at the silica surface, which leads to a reduction in silica solubility. In the absence of cations that form insoluble silicates, the solubility of amorphous silica increases as the solution pH exceeds 8 because of increased dissociation of H_4SiO_4 into the bisilicate anions $H_3SiO_4^-$ and $H_2SiO_4^{2-}$ (Iler, 1979).

If the silica solubility is exceeded either by a pH change or a decrease in temperature, the excess monosilicic acid will polymerize or condense to form colloid particles (known as colloidal silica), as small as 10-20 Å in size, by a series of reactions involving linear, cyclic and three-dimensional silicic acids (Alexander et al., 1953; Iler, 1979). Silica polymerisation in water typically follows first or second order kinetics, although reaction orders between 1 to

8 have previously been reported. The rate of polymerisation is generally found to increase with supersaturation and temperature (up to ~100°C). The polymerisation is catalysed by hydroxyl ions in slightly alkaline or neutral solutions to reach a maximum between pH 6-9. At pH values above 9, the solubility of silica rapidly increases thereby reducing the extent to which polymerisation can take place. Salinity is also expected to affect the rate of silica polymerisation because of a reduction in silica solubility with increasing salt concentrations (Chan, 1989; Rothbaum and Rohde, 1979).

Table 1.2 Solubility of amorphous SiO_2 in water at pH < 9.0 (Krumbein and Werner, 1983)

Temperature (°C)	SiO_2 solubility (ppm)
25	100-140
60	190-250
90	300-380

Apart from the polymerization of multiple units of SiO_2, colloidal silica species can also be formed between silica and other organic or inorganic compounds. For example, silica is commonly found to co-precipitate with magnesium and aluminum hydroxides by complexation and adsorption in biological, geological and industrial systems (Ning, 2002; Perry and Keeling-Tucker, 2000).

1.2 Composite Fouling in Cane Sugar Mill Evaporators

The evaporator system in sugar mill is a multi-effect evaporator train that increases the sugar concentration of the cane juice from approximately 12 % to 60-65%. Evaporation systems usually consist of four to five stages or effects designed to appropriately distribute the evaporation load and to maximise the steam economy. For a quadruple effect system, the steam economy can be as high as ~3.6 tons of water evaporated for 1 ton of vapour fed (total vapour consumption of 32% of cane). Such a high economy is due to water evaporated from one effect acting as the steam supply for the next effect, and each subsequent effect uses a lower temperature and pressure to maintain the thermal driving force of the system (Honig, 1963) (Fig 1.1).

Multiple effect evaporation of cane juice leads to the accumulation of organic and inorganic non-sugar scales on the surface of calandria tubes in sugar mill evaporators. The scales that are formed reduce the heat transfer efficiency and the capacity of the evaporator station. Chemical cleaning of the evaporator must be undertaken every 7 to 17 days, and not only is costly but also adds to effluent discharges, thereby posing environmental problems. Additional costs of evaporator scaling are associated with reduced sugar quality from extended residence time of juice during evaporation and equipment corrosion (Doherty, 2000; Walthew and Turner, 1995; Walthew et al., 1997).

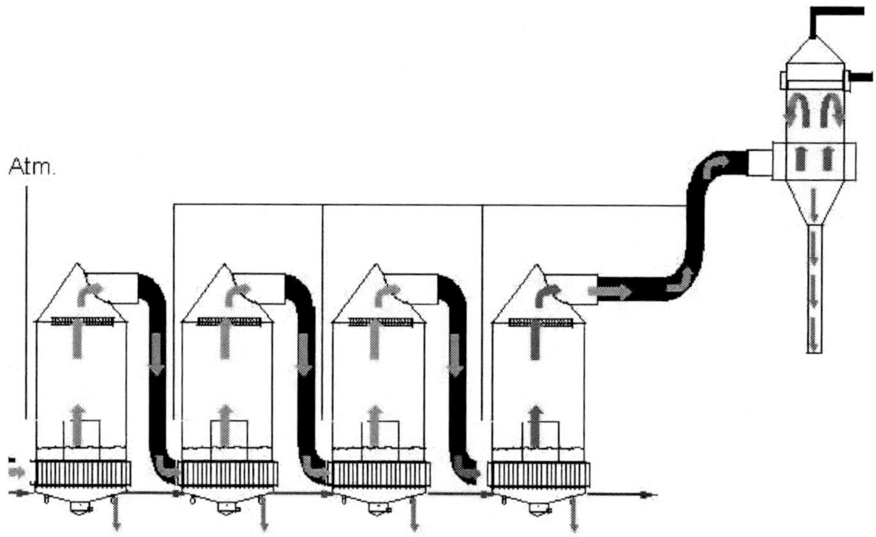

Fig. 1.1 Quadruple effect evaporator set of a sugar mill

The composition of evaporator scale varies depending on the stage of evaporation cycle of the system and may include phosphates, silica and organic acid salts (Nandagopal and Ramamurthy, 1976; Walthew, 1996). Phosphate scales are more commonly observed in the earlier effects while silica and calcium salts of organic acids such as oxalate in either one of its crystalline forms, i.e., COM and COD (see Table 1.3), usually deposit in the latter vessels (i.e., 4th and 5th effects) because of temperature and concentration effects during juice evaporation. The solubilities of calcium oxalate and silica decrease with increasing concentration of sugar (Walford and Walthew, 1996; Walthew et al., 1998). Unlike most other scalants, calcium oxalate and amorphous silica (SiO_2) form a hard composite scale intractable to conventional cleaning methods and chelating agent EDTA is often required for its effective removal from the vessel (Doherty, 2000).

Table 1.3 A typical scale composition in the fifth vessel of a multi-effect evaporator station in an Australian sugar mill (Doherty, 2000)

Component	Concentration (wt %)
Calcium oxalate (as COM and COD)	51.3
Silica	33.3
Hydroxyapatite	7.8
Others	7.6

The precipitation of oxalate is either due to the existence of oxalic acid in the sugar cane plant or decomposition of other components of the juice to oxalic acid during processing. Silica is an essential trace element in the cane plant tissue and is present in the dirt carried into the sugar factory with the cane supply. Calcium is introduced into the mill via cane plant and by lime ($Ca(OH)_2$) addition during juice clarification (Poynton and Alexander, 1957; Walford and Walthew, 1996).

No information is available in the literature regarding the co-precipitation of COM and SiO_2. Doherty et al. (1995) investigated the inhibition of COM crystallization by polymeric additives such as malic acid copolymer and polyacrylic acid. The results indicated that the interference of polymer molecules during the nucleation and crystal growth of COM precipitation extended the induction periods and modified the crystal morphology of COM. The inhibitory efficiency of the additives in the synthetic juice solutions were found to be lower than those in the pure solution due to the competitive adsorption at growth sites of COM crystal surface by other ionic species.

Walthew et al. (1998) demonstrated that silica concentration in clear juice could be reduced by extracting sugar plant under controlled pH conditions. However, pH control during extraction also reduced lime addition giving rise to other processing problems such as equipment corrosion and sucrose inversion; it also had little influence on the silica fouling in the final effect of evaporator and so would have limited applications in sugar factory.

1.3 Objectives of Present Work

The overall aim of this project was to study the mechanisms of composite fouling of SiO_2 and COM in cane sugar evaporators. The specific objectives of this work were as follows:

i. to investigate the possible COM-SiO_2 interactions occurring in water and sugar media, and the effect of these interactions on the kinetics and thermodynamics of co-precipitation of COM and SiO_2 and the characteristics of composite deposits;
ii. to determine the subcooled nucleate flow boiling heat transfer coefficients of water and sugar solutions under a range of thermal hydraulic operating conditions;
iii. to compare the measured heat transfer coefficients with the predictions of published flow boiling models and to propose a model applicable to this work;
iv. to evaluate and quantify the effects of operating conditions such as solution composition, velocity and temperature on composite fouling behaviour in synthetic juice solutions;
v. to determine factors governing the composite scale formation; and
vi. to develop a composite fouling model to predict fouling propensity in cane sugar evaporators based on the mechanistic assumptions proposed.

Chapter 2

THEORETICAL BACKGROUND

2.1 HEAT EXCHANGER FOULING THEORIES

2.1.1 Effects of Fouling

Fouling, a term originally used in the oil industry, is defined as the accumulation of deposits on heat exchanger surfaces, which obstructs the transfer of heat and increases the resistance to fluid flow (Epstein, 1983; Taborek et al., 1972b). The growth of these deposits causes the thermal and hydrodynamic performance of heat transfer equipment to decline with time. Fouling-related costs for an industrial process using heat transfer equipment may include (Bohnet, 1987; Bott, 1995; Muller-Steinhagen, 1999):

- Increased capital investment, e.g., due to more materials for the construction of heat transfer equipment;
- Additional operational cost, e.g., higher energy consumption due to thermal inefficiencies and pressure drop;
- Higher maintenance cost, e.g., due to local corrosion of tubes and additional cleaning systems;
- Loss of production, e.g., due to deteriorating product quality or inability to meet operational demands, and enforced shutdown for cleanup;
- The cost of remedial action, such as additives to reduce fouling and chemicals to clean fouled surfaces.

It is estimated that the total cost of heat exchanger fouling accounts for about 0.25% of the GDP of industrialized countries during the 1980s and early 1990s (Muller-Stainhagen, et al., 2005).

2.1.2 Types of Fouling

Fouling is generally classified into five categories according to the principal process which gives rise to the phenomenon: crystallization, particulate, chemical reaction, corrosion and biological fouling (Epstein, 1981b; Taborek et al., 1972b).

Crystallization Fouling

This involves the crystallization of dissolved species from solution onto the heat exchange surface when process conditions lead to supersaturation of the dissolved inorganic salts at the heat transfer interface, which is often referred to as scaling. It also occurs when deposits are formed on heat transfer surfaces (HTS) by freezing a liquid or a high-melting point constituent of a multi-component fluid onto the subcooled surface, which is normally referred to as solidification or freezing fouling.

Particulate Fouling

Particulate fouling is defined as the deposition of particles suspended in liquid onto HTS. Suspended particles may include pollutants/species always present in the feed stream or products of the chemical reactions occurring within the fluid.

Chemical Reaction Fouling

It includes deposits that are formed at the HTS as a result of chemical reaction within the process fluid. The HTS is not a reactant, although it may act as a catalyst.

Corrosion Fouling

In this case, the heat exchanger material reacts with the fluid to form corrosion products on the HTS. This classification is restricted to in situ corrosion processes. Ex situ corrosion may result in either precipitation or particulate fouling, depending on whether corrosion products are soluble or insoluble at the process fluid conditions.

Biological Fouling

Biofouling refers to the development and deposition of organic films consisting of micro-organisms and the attachment and growth of macro-organisms.

2.1.3 Sequence of Fouling

The overall fouling process can be viewed as consisting of the following five sub-processes (Epstein, 1981a):

Initiation

Conditions that promote subsequent fouling are established during the initiation period, e.g., nucleation sites for crystallization or nutrients for biological growth are developed.

Transport to Surface

Mechanisms that contribute to foulant migration to the HTS include diffusiophoresis, turbulent diffusion, thermophoresis, brownian diffusion, electrophoresis, gravity, etc.

Attachment

Factors that may influence the adhesion of foulant are van der Waals forces, electrostatic forces and external force fields at wall.

Removal

Removal of foulant may occur as a result of spallation, erosion or dissolution, which may or may not begin soon after initial layer is deposited.

Aging

Aging of the deposit starts as soon as it has been attached on the HTS. The process of aging includes changes in the crystal and chemical structures of foulant by dehydration, polymerization, developing thermal stress, etc., which may affect the strength of the deposit and hence the removal process. Aging is the least investigated step and is usually ignored in the modeling studies.

2.1.4 Fouling Curves

The buildup of the fouling deposits as a function of time may be regarded as the result of deposition accompanied by a removal process, which may form one of the four curves shown in Fig 2.1 (Bohnet, 1987; Knudsen and Roy, 1982; Muller-Steinhagen, 1999) depending on the dominant fouling steps:

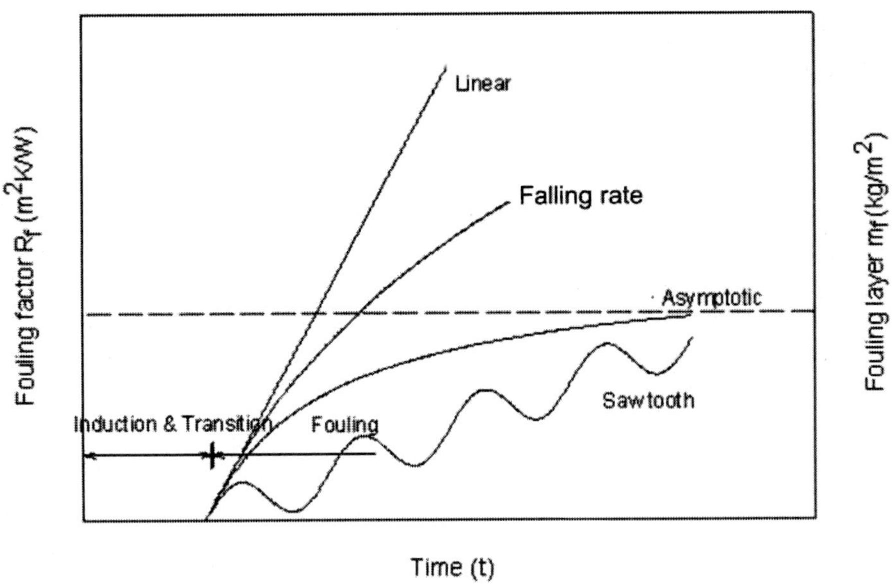

Fig. 2.1 Different fouling curves

- Linear curve: the fouling resistance increases as linear function of time. This is observed for very hard scales with high adhesion and indicates constant deposition and removal rates (or no removal).
- Decreasing rate curve: the fouling resistance also increases with time but at a decreasing rate. This behavior is found for scales with low mechanical strength and results from a falling deposition rate and/or increasing removal rate.

- Asymptotic curve: the fouling resistance eventually reaches a constant value. It is obtained if the deposition rate remains constant while the removal rate increases with the thickness of the deposit and eventually equals the deposition rate.
- Saw tooth curve: this is observed when the deposited material is rapidly removed due to periodic changes of the conditions.

2.1.5 Conditions that Influence Fouling

Conditions that could influence fouling or that affect the impact of fouling on heat exchange or fluid flow can be divided as operating parameters, fluid properties and heat exchanger design parameters (Garrett-Price et al., 1985).

2.1.5.1 Operating Parameters

Surface temperature, bulk temperature and fluid velocity are among the most important parameters affecting fouling behaviour, e.g., the rate of fouling. For a surface adhesion controlled fouling process involving chemical reactions, e.g., crystallization or polymerization, higher surface temperature tends to increase fouling rate due to an increased reaction rate and, in the case of inverse solubility salts, a reduction in the solubility of compounds. Also, variations in the fluid viscosity between the heat transfer surface and bulk fluid may affect transport of fouling species due to temperature gradients across the wall boundary layer (Bott, 1997).

Fluid velocity is a critical factor with respect to both deposition and removal processes. Higher velocities will enhance the mass transfer of fouling species from the bulk fluid to the heat transfer surface and the intensity of the removal forces, but tend to reduce the efficiency of surface adhesion for foulant. Thus the deposition rate may increase or decrease with velocity depending on the specific mechanisms controlling deposition process (Bott, 1997; Pinheiro, 1986; Yiantsios and Karabelas, 2003).

2.1.5.2 Fluid Properties

Both the nature of the fluid and the nature of the species dissolved or entrained in the fluid are known to affect fouling processes. For example, water quality in cooling water systems has a significant impact on potential fouling mechanisms, which may involve crystallization of inversely solubility salts, deposition of particulate matters, corrosion and biological fouling (Knudsen, 1981).

2.1.5.3 Heat Exchanger Parameters

Heat exchanger parameters such as surface materials, surface structure and type and geometry of exchanger could also affect the fouling potential. Rough surfaces, for instance, are known to promote fouling by providing additional nucleation sites for deposition (Knudsen and Roy, 1982; Suitor et al., 1977). Copper and brass alloy surfaces generally foul more than stainless steel surface from the added effect of corrosion (Suiter, et al., 1977; Somerscales, 1999) and higher surface energy (Tyson, 1975; Mantel, et al., 1995). Plate heat exchangers also tend to have lower fouling rates than tubular exchangers due to the turbulence generated by the corrugations on the plate (Knudsen and Roy, 1982).

2.1.6 Fouling Resistance

For steady state conditions, the heat flux, q", across a clean surface is given as:

$$q" = U_C \Delta T_m \tag{2.1}$$

where ΔT_m is the log mean temperature difference between hot and cold fluids and U_C is the overall heat transfer coefficient (based on hold fluid side heat transfer area, A_h) for clean conditions, i.e.,

$$\frac{1}{U_C} = R_{TC} = \frac{1}{h_h} + R_W + \frac{A_h}{A_c h_c} \tag{2.2}$$

where R_{TC} is the total resistance to heat flow, A_c is the cold fluid side heat transfer area, $1/h_h$ and $A_h/A_c h_c$ represent the film resistance of the hot and cold fluids, respectively. R_W is the resistance to heat flow of the metal wall.

The heat flux across a fouled surface is given as:

$$q = U_D \Delta T_m \tag{2.3}$$

Where

$$\frac{1}{U_D} = R_{TF} = \frac{1}{h_h} + R_W + R_f + \frac{1}{h_c} \tag{2.4}$$

where R_f is the resistance of fouling to heat transfer. Thus the so-called fouling resistance or fouling factor is defined as:

$$R_f = \frac{1}{U_D} - \frac{1}{U_C} \tag{2.5}$$

and

$$q" = \frac{1}{\frac{1}{U_C} + R_f}(T_w - T_b) \tag{2.6}$$

where T_w and T_b are wall and bulk temperatures, respectively. If the wall temperature T_w is to be kept constant, heat flux q" decreases with the increasing R_f (and consequently results in a lower temperature of cold fluid). Alternatively, if the heat flux q" remains constant, i.e., the deposit/fluid interface temperature remains the same as the clean wall temperature value, T_W must be increased with the increasing R_f (Knudsen and Roy, 1982).

If the total available heat transfer area A is taken into account, the rate of heat transfer q is expressed as:

$$q = U_D A \Delta T_m \tag{2.7}$$

In the design of heat exchanger, it is necessary to determine the area (A) required for heat transfer. So rearranging the above equation provides a means of estimating the required heat transfer area under fouling conditions and the impact of fouling on heat exchanger design:

$$A = \frac{q}{U_D \Delta T_m} = (\frac{1}{U_C} + R_f)\frac{q}{\Delta T_m} \tag{2.8}$$

As seen in eqn (2.8), the choice of the individual fouling resistance for the calculation of U_D can have a marked influence on the size of the heat exchanger and hence the capital cost.

While the fouling resistance for a gas/gas heat exchanger represents a negligible additional capital cost due to the low thermal conductivity of gases, the same choice of fouling resistance represents a considerable increase in the required surface for the liquid/liquid heat exchanger. Garrett-Price et al. (1985) and Muller-Steinhagen (1999) found that heat exchangers built between 1970-1990 are in average oversized by 35-38% to allow for fouling.

2.1.7 Composite Fouling

Composite fouling is a form of fouling process involving more than one type of fouling species or mechanism, which may interact with each other to produce the final deposit. There is a knowledge gap associated with a lack of understanding of the mechanisms underlying the composite fouling process despite its practical significance. Conventional study to date on fouling mechanisms has been restricted to single-component systems and only in a limited number of cases have the effect of interfering species on the fouling process of interest been reported (Oliveira et al., 1993; Sheikholeslami et al., 2001; Sudmalis and Sheikholeslami, 2000). It is known that the co-existing compounds often interact to give rise to effects that may be synergistic or antagonistic, the results of which are unpredictable on the basis of single component studies (Sheikholeslami and Ng, 2001). Also, industrial process streams such as those in the sugar mill evaporators very often contain different types of fouling species, which may impact on the removal of the composite scales by altering their strength and structure. This has necessitated the need to study the interactive effects in multi-component fouling systems in order to provide a better understanding of the basics of composite fouling mechanisms encountered in sugar mill evaporators.

Sudmalis and Sheikholeslami (2000) studied the co-precipitation of calcium salts of carbonate and sulfate. The precipitation kinetics and thermodynamics of both $CaCO_3$ and $CaSO_4$ were affected by the co-precipitation, which also led to changes in the structure and strength of resulting deposits. Oliveira et al. (1993) investigated the effect of kaolin particles on biofouling of microorganism, *pseudomonas fluorescens*. The results from dynamic fouling experiments indicated that biofilm growth in bacteria-kaolin mixtures was higher than that in the pure bacteria suspensions, especially at higher velocities. Andritsos and Karabelas (1999) examined calcium carbonate scale formation in the presence of silica particles. The authors showed that particulate silica deposition led to a reduction in the mean calcite crystal size formed on the metal surface, although it had a negligible contribution to the total mass of deposition.

On the other hand, the reduced crystallization fouling rate of calcium sulphate on a plate heat exchanger was achieved by addition of non-crystalline alumina particles in solution (Bansal et al., 1997). Particles attached on the crystals reduced the deposit strength and inhibited crystal growth by acting as distorting agent. Similar inhibitive behaviour was also obtained for the particulate fouling of kaolin suspensions in the presence of magnetite particles. The kaolin-magnetite mixtures were found to have a lower fouling rate than pure kaolin solution due to the surface interactions between the two types of particles, which inhibited the deposit growth (Oliveira et al., 1993).

2.1.8 Effect of Boiling on Fouling

The presence of boiling on heat transfer surface has a significant impact on fouling rate and deposit characteristics in various industrial applications including chemical process evaporators, desalination units and steam generators. The main source of fouling species in boiling water systems are dissolved inorganic salts (e.g., calcium carbonate and calcium sulphate) and suspended particles and corrosion products (e.g., iron oxide) (Epstein, 1981a; Gudmundsson, 1981; Melo, 1988).

Fouling is generally enhanced by boiling heat transfer due to the high temperatures and additional bubble activities involved at heating surface. Partridge and White (1929a) investigated the boiler scale formation by calcium sulphate. The deposit was shown to exhibit characteristic ring structure typical of those formed as a result of microlayer evaporation at the base of bubble nucleating sites. These results were later confirmed by Freeborn and Lewis (1962) and Hospeti and Mesler (1965). Asakura et al. (1978) developed a particulate fouling model based on the concept of microlayer evaporation to describe the deposition of iron oxide particles in boiling water. Good agreement between the measured deposition rate coefficients and model predictions was reported (see section 4.8 for further details). Helalizadeh et al. (2005) examined the calcium sulphate-calcium carbonate mixed salt precipitation under convective heat transfer and subcooled flow boiling conditions, and developed a mechanistic model based on single phase forced convection data and a boiling enhancement factor. The new model was able to predict the rate of mixed salt precipitation fouling under flow boiling conditions reasonably well with deviations ranging from 6-25%.

The effect of boiling on the particulate fouling of iron oxide was studied by Charlesworth (1970). The deposition rate on boiling surface was found to be an order of magnitude larger than that for the unheated surface. The author suggested that the higher fouling rate in the presence of boiling might be caused by fresh water flux to the surface of the deposit layer to replace the departing bubbles. The iron oxide deposit formed under boiling condition was porous in nature probably because of bubble nucleation within the deposit. Klimas et al. (2004) investigated fouling enhancement by colloidal iron oxides under flow boiling conditions. An increase in fouling rates by up to 3 orders of magnitude was observed upon reaching a certain elevated steam quality and mixture velocity.

The rates of crystallization, corrosion and particulate fouling under boiling conditions were strongly dependent on heat flux due to its effect on the rate of bubble generation. The fouling rate depended on heat flux to the power of 1 to 5.5 with the most common values between 1-2 (Charlesworth, 1970; Epstein, 1981a; Somerscales, 1992).

2.2 CRYSTALLIZATION

2.2.1 Definition

Crystallization is generally considered a process of generating a solid crystalline material from a supersaturated solution. It is classified as a diffusional operation accompanied by formation of a new heterogeneous phase (Long, 1995; Mullin, 1979).

2.2.2 The Crystalline State

Solid may be crystalline or amorphous, and the crystalline state differs from the amorphous state in the regular arrangement of the constituent molecules, atoms or ions into some fixed and rigid pattern known as a lattice. The term of "crystalline" is most frequently used to indicate a high degree of internal regularity, resulting in the development of definite external crystal faces, although many of the substances that were once considered to be amorphous have now been shown (by X-ray analysis) to exhibit some degree of regular molecular arrangement (Mullin, 1997).

The regularity of the internal structure of the solid body results in a characteristic shape of the crystal, with the planes of crystal faces being parallel to atomic planes in the lattice. However, rarely do any two crystals of a given substance look identical. This is especially true for the natural minerals grown under different restraining conditions, which result in stunted growth in one direction and exaggerated growth in another (Mullin, 1997).

A substance capable of crystallizing into different, but chemically identical, crystalline forms is said to exhibit polymorphism as a result of changes in the crystallization conditions such as temperature and solvent. The commonly known examples of trimorphous substance include calcium carbonate and calcium oxalate (e.g., calcium oxalate monohydrate, dihydrate and trihydrate, see section 1.1.1) (Mullin, 1997).

2.2.3 Solubility

The solubility of a sparingly soluble salt, designated here as MA, in water can be expressed in terms of the concentration solubility product (or solubility constant), K_c ($(mol/L)^2$), according to the following dissociation equilibrium if one molecule of MA dissociates into x positive ions and y negative ions:

$$M_xA_y \Leftrightarrow xM^{z+} + yA^{z-} \tag{2.9}$$

$$K_c = (M^{z+})^x (A^{z-})^y \tag{2.10}$$

Where z+ and z- are the valencies of the ions. For salts that produce two ions per molecule (i.e., x=y=1), the solubility, c* (mol/L), is obtained by:

$$c^* = (K_c)^{1/2} \tag{2.11a}$$

In general (Mullin, 1997),

$$c^* = (K_c / x^x y^y)^{1/(x+y)} \tag{2.11b}$$

It should be noted that the use of the simple solubility product principle is limited to very dilute aqueous solutions, i.e., for concentration up to ~10^{-3} mol/L in practice, beyond which a more fundamental approach involving activity concepts should be used (Atkins, 1994; Mullin, 1997).

The solubility of a sparingly soluble salt in water may be influenced by a number of factors including solution pH and temperature, the formation of metastable crystal phase on the surface of a bulk crystal and the presence of other ionic/polymeric species in solution. The latter two factors are of particular importance in determining the solubility of calcium oxalate in various biological and industrial processes (Furedi-Milhofer et al., 1987; Robertson, 1982; Ulmgren and Radestrom, 1999a).

2.2.4 Supersaturation

Supersaturation is defined as the amount of solute concentration in excess of equilibrium solubility for a given temperature and is the most important parameter that defines the crystallization process. There are many methods to induce supersaturation in a solution, e.g., by removing some of the solvent from the solution (through evaporation or freezing), cooling/heating the solution (for normal/inverse solubility salts), or adding another component as a precipitating agent to the reaction environment, thus reducing the solubility of the solute (Dirksen and Ring, 1991; Mullin, 1979). One example of this is reducing solubility of calcium oxalate by adding magnesium ions (Ulmgren and Radestrom, 1999b).

The supersaturation of a system may be expressed in a number of different ways. Among the most common expressions of supersaturation are concentration driving force, Δc (= c – c*), the supersaturation ratio, SS (= c/c*), and the relative supersaturation ratio, α_r (Δc/c* = SS-1), where c is the solution concentration and c* is the equilibrium value at the given temperature (Mullin, 1997) (also see section 4.1.1.1).

The fundamental driving force for crystallization is the difference between the chemical potential of the given substance for supersaturated solution, μ_1, and the solid crystal face μ_2, i.e.,

$$\Delta\mu = \mu_1 - \mu_2 \tag{2.12}$$

The chemical potential, μ, is defined as:

$$\mu = \mu_0 + RTIn(a) \tag{2.13}$$

where μ_o is the standard potential, a is the solute activity, R is the gas constant and T is the absolute temperature. The dimensionless thermodynamic driving force may therefore be expressed as:

$$\frac{\Delta\mu}{RT} = n\ln(a / a^*) = n\ln(fc / f^* c^*) = n\ln(SS) \tag{2.14}$$

where a^* is the activity of a saturated solution (i.e., at crystal), n is the number of ions in a molecular unit, f and f^* are the activity coefficients of solute in solution and crystal, respectively, and $SS = a/a^*$ represents the definition of dimensionless supersaturation ratio. The above equation may be expressed in terms of the relative supersaturation, α_r:

$$\frac{\Delta\mu}{RT} = n\ln(\alpha_r + 1) \approx n\alpha_r \tag{2.15}$$

where α_r is much less than unity ($\alpha_r < 0.1$) for low supersaturations. In practical crystallization calculations, supersaturation ratio is mostly expressed in terms of concentrations by setting f and f^* as unity (Mullin, 1997; Sohnel and Garside, 1992) because of its ease of use from experimental measurements, i.e.,

$$\alpha = \frac{c - c^*}{c^*} \tag{2.16}$$

2.2.5 Nucleation

The condition of supersaturation alone is not sufficient cause for a system to begin to crystallize. There must exist in the solution a number of minute solid bodies, embryos, nuclei or seeds that act as centers of the crystallization (Garside, 1982).

The nucleation process can be divided into two main categories: primary and secondary. Primary nucleation refers to all cases of nucleation through excess chemical potential. Primary nucleation can be either homogeneous or heterogeneous depending on whether the new crystals are generated in the absence of or in the presence of other crystals, foreign particles or surfaces. Heterogeneous nucleation is generally much more common than homogeneous nucleation. Secondary nucleation occurs when new nuclei are formed in the vicinity of crystals already present in the supersaturated systems (Mullin, 1979; 1997).

The process of homogeneous nucleation is assumed to occur by a series of bimolecular association resulting in ordered aggregates, clusters or embryos. Some embryos grow to nuclei, whose size is the critical size r_c, by additional collisions, but others will disappear (disperse) by decreasing their size or contributing to the growth of crystals in supersaturation (Garside, 1982).

Supersaturated solutions have a population of subcritical sized molecular clusters which form and dissolve. Nucleation occurs when one of these clusters exceeds the critical size. The number of molecules in a stable crystal nucleus can vary from about ten to several thousands (Mullin, 1997). It was estimated that that the average cluster size in supersaturated solution of glycine was of the order of two molecules. While the average cluster size was small, large clusters of hundreds of molecules existed, but only there were a few of these. Most of the molecular association was in the form of dimers and trimers (Myerson, 1993).

According to the classical theories of primary homogeneous nucleation, stemming from the work of Gibbs, Volmer and others, the overall free energy, ΔG, of the embryo is the sum of two terms: the free energy due to formation of a new volume (ΔG_V) and the free energy due to the new surface (ΔG_s) created:

$$\Delta G = \Delta G_s + \Delta G_V \tag{2.17}$$

Assuming that the embryos are spherical of radius r, it follows that:

$$\Delta G = 4\pi r^2 \gamma_{12} + \frac{4}{3}\pi r^3 \Delta G_v \tag{2.18}$$

where ΔG_v is the free energy change of the transformation per unit volume and γ_{12} is the interfacial tension (energy) between the crystallization medium, phase 1, and the nucleating cluster, phase 2. Due to the competition between the volume and surface term, the free energy of formation, ΔG, passes through a maximum at certain value of r^* (Mullin, 1997).

The critical nuclei size r^*, can be obtained by minimizing the free energy function with respect to radius, i.e., $d\Delta G/dr = 0$, giving:

$$r^* = \frac{-2\gamma_{12}}{\Delta G_v} \tag{2.19}$$

The corresponding change in the free energy associated with the formation of this critical nucleus, ΔG_{crit}, is the activation energy for nucleation, which has to be surmounted before the final equilibrium state is attained:

$$\Delta G_{crit} = \frac{4\pi\gamma_{12} r^{*2}}{3} \tag{2.20}$$

The behavior of newly created crystalline lattice structure in a supersaturated solution depends on its size; it either grow or redissolve, but the process which it undergoes should result in the decrease in the free energy of the particle. The critical size r^*, therefore, represents the minimum size of a stable nucleus (Garside, 1982; Mullin, 1997).

The basic Gibbs-Thomson relationship describing the relationship between the solubility of a crystal and its size may be written as:

$$In(\frac{c}{c*}) = In(SS) = \frac{2\gamma_{12}V}{kTr} \qquad (2.21)$$

Where V is the molecular volume, k is the Boltzmann constant (=R/N, R is the gas constant = 8.314 J/K mol, and N = the Avogadro number = $6.023x10^{23}$ mol^{-1}). Incorporating the above equation into the free energy change equations gives

$$-\Delta G_v = \frac{2\gamma_{12}}{r} = \frac{kT\ln(SS)}{V} \qquad (2.22)$$

$$\Delta G_{crit} = \frac{16\pi\gamma_{12}^3V^2}{3[kT\ln(SS)]^2} \qquad (2.23)$$

The rate of nucleation, J, can be expressed in the form of the Arrhenius law commonly used for determining the rate of a thermally activated process:

$$J = A\exp(-\Delta G / kT) \qquad (2.24)$$

Or written as

$$J = A\exp\left[-\frac{16\pi\gamma_{12}^3V^2}{3k^3T^3[\ln(SS)]^2}\right] \qquad (2.25)$$

This equation indicates three main variables that govern the rate of nucleation: temperature T; degree of supersaturation, SS; and interfacial tension, γ_{12} (Garside, 1982; Mullin, 1997).

Since heterogeneous nucleation can occur at lower degrees of supersaturation than homogeneous nucleation, the overall free energy change associated with the formation of a critical nucleus under heterogeneous conditions must be less than that of homogeneous nucleation, i.e.,

$$\Delta G'_{crit} = \phi\Delta G_{crit} \qquad (2.27)$$

where ϕ<1 (Melia, 1965).

2.2.6 Induction Period

Induction period is the time that elapses between the establishment of supersaturation and the detection of crystals. This time lag is considerably influenced by the level of supersaturation, the state of agitation, the presence of impurities, etc. (Garside, 1982; Mullin, 1997; Sohnel and Garside, 1992). Although the induction period is often assumed to be a measure of nucleation rate, it is not, by itself, a fundamental characteristic of a crystallizing system due to the fact that a different result can be obtained depending on experimental techniques employed (e.g., by visual measurement or by laser light scattering measurement) (Garside, 1982; Mullin, 1997).

2.2.7 Ostwald's Rule of Stages

Based on the experimental observations that some inorganic salt solutions, when cooled rapidly, first deposited crystals of a less stable form than that which normally crystallizes, Ostwald made a generalized statement in 1897 as a so-called rule of stages. It states that an unstable system does not necessarily transform directly into the most stable state, but is preceded by one which most closely resembles its own, i.e., into another transient state which possesses greater energy than that of the stable phase (Mullin, 1997; Sohnel and Garside, 1992).

Ostwald's law is confirmed by the behaviour of many crystallization systems including those of calcium phosphate and calcium carbonate. However, there have been many exceptions to this rule there thermodynamic explanations alone do not offer adequate theoretical support. Instead, a combined thermodynamics-kinetics approach does appear to provide some justification. In many cases the most probable explanation of the phenomenon lies in the kinetics of the transformation, the deciding factor being the relative rates of crystal nucleation and growth of the more-stable and less-stable forms. The resulting reaction is not the one that is thermodynamically possible, but the one that has the fastest rate. In other words, kinetic factors are often more important than thermodynamic factors (Mullin, 1997; Sohnel and Garside, 1992). More details on the polymorphic formation of calcium oxalate as a result of this law are provided in section 4.3.

2.2.8 Crystal Growth

Mechanisms of crystal growth are much more studied compared with those of nucleation. The process of crystal growth occurs mainly in the following steps: i) transport of solute from the bulk solution to the crystal surface; ii) adsorption onto the crystal surface; iii) diffusion over the surface to a growth site and iv) integration into the crystal lattice. Desolvation and release of heat of crystallization can also take place during and after steps i-iv. The rate-limiting step can be any of the above depending on growth conditions including supersaturation, temperature, additives, solvent and hydrodynamics of the system (Dirksen and Ring, 1991; Mullin, 1979). A number of theories are generally available to explain the growth of single crystals from solution. They include adsorption layer theories and those based on diffusion-integration (Mullin, 1997; Ohtaki, 1998).

2.2.8.1 Adsorption Layer Theories

The adsorption layer theory, or the so-called Gibbs-Volmer theory, suggests that crystal growth is a discontinuous process, taking place by adsorption where layer upon layer on the crystal surface is responsible for growth. A certain size of cluster called monolayer island nucleus is usually created before every new layer commences to build up. Adsorbed growth units either join an existing surface nucleus or form a new nucleus, which further grows through the addition of other growth units. The two-dimensional nucleus then spreads across the surface to form a monomolecular layer (Mullin, 1997). A detailed theoretical treatment of crystal growth is given by Kossel, which assumed the stepwise growth of a mononuclear layer to take place on a surface of a perfect crystal. The results indicate that a reasonably high degree of local supersaturation is necessary for two-dimensional nucleation to occur, but lower than that required for the formation of three-dimensional nuclei under equivalent conditions. Two-dimensional nucleation is unfavourable in a low supersaturated solution and this process becomes the rate-determining step of the crystal growth. Therefore, the theory predicts that the crystal growth can not occur under the condition of a very low supersaturation ratio, which is contrary to experimental observations (Mullin 1997, Ohtaki 1998). This dilemma of the two-dimensional nucleation mechanism can be cleared up by the screw dislocation mechanism proposed by Burton et al (1951).

The theory of screw dislocation states that few crystals ever grow in the ideal layer-by-layer fashion without some imperfection, i.e., screw dislocations, occurring in the pattern, which cause steps to be formed. The dislocation proceeds in a spiral form on the propagation of the step. This mechanism enables the vertical growth of crystal continuously without the interruption of the two-dimensional nucleation. It also successfully explained the rate of crystal growth at a low degree of supersaturation. The screw dislocation proposed has been observed on the surface of many kinds of crystals of actual systems and supported by computer simulations (Mullin, 1997).

2.2.8.2 Diffusion – Reaction Theory

The diffusion-integration concept involves the assumption that material is deposited on the crystal surface at a rate proportional to the difference in concentration between the point of deposition and the bulk solution. Two basic steps of crystal growth (Fig 2.2) are given by the diffusion-integration theory: diffusional process whereby solute is transported from the bulk fluid phase through the solution boundary layer close to the crystal surface, followed by integration of adsorbed solute ions or molecules at the crystal surface into the crystal surface (first order reaction when the solute molecules arrange themselves into the crystal lattice) (Mullin, 1979; Mullin, 1997).

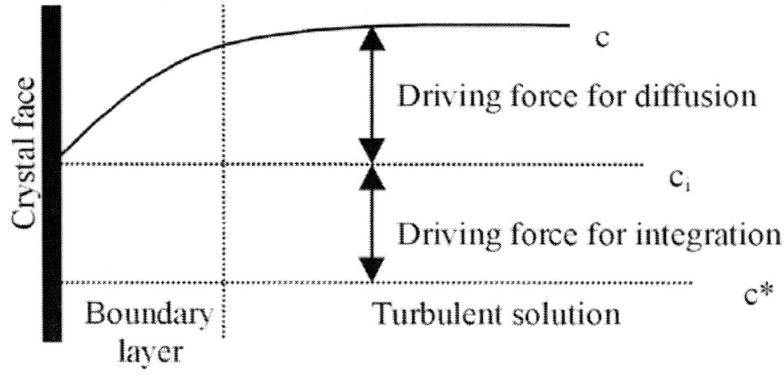

Fig. 2.2 Concentration driving force for crystal growth from solution

These two steps, which proceed under the influence of different driving forces, can be represented by the following equations (assuming a first order reaction):

$$\frac{dm}{dt} = K_d A(c - c_i) \tag{2.28}$$

$$\frac{dm}{dt} = K_r A(c_i - c^*) \tag{2.29}$$

where m = mass deposited in time t; A = crystal surface area; c, c_i and c^* are solute concentrations in bulk solution, at the interface and at equilibrium saturation. K_d and K_r are diffusion and integration mass transfer coefficients.

It is usually convenient to eliminate c_i, which is difficult, if not impossible, to measure, by considering the overall concentration driving force $\Delta c = c\text{-}c^*$. A generalized equation for crystallization based on this overall driving force can be written as:

$$\frac{dm}{dt} = K_G A(\Delta c)^n \tag{2.30}$$

where K_G is an overall crystal growth coefficient, n refers to as the order of the overall crystal growth process. Unlike its relevance when used in chemical kinetics, n has no fundamental significance and can not give any indication of the number of elementary species involved in the growth process (Mullin, 1997).

If $n = 1$, the interfacial concentration, c_i, may be eliminated from the equations to give

$$K_G = \frac{K_d K_r}{K_d + K_r} \tag{2.31}$$

For extremely rapid reaction, i.e., large K_r, $K_G = K_d$, and the crystallization is controlled by the diffusional operation and vice versa. The diffusional step is generally considered to be linearly dependent on the concentration driving force, but the validity of the assumption of a first-order surface reaction is highly questionable. Many inorganic salts crystallizing from aqueous solution give an overall growth rate order, n, in the range of 1 – 2 (Mullin, 1979).

A complicating factor in using diffusion-controlled growth rate expressions is the fact that very small crystals can have solubilities significantly higher than those of macrocrystsls (Gibbs-Thomson effect). In any complete analysis of the growth process, therefore, the combined effects of diffusion, surface integration and size-solubility may have to be considered together. Although the diffusion theories of crystal growth have grave deficiencies (e.g., an inability to explain layer growth or the faceting of crystals), crystal growth rates are conveniently measured and reported in diffusional terms. The utilization of the mathematics of mass transfer processes also makes it the preferred approach despite its many limitations (Mullin, 1997).

2.2.9 Crystal Habit Modification

The external appearance, or habit, of a crystal can be classified into seven systems according to the types and combination of crystal symmetries it possesses: regular, tetragonal, orthorhombic, monoclinic, triclinic, trigonal and hexagonal. The crystal system favored by a compound reflects to some extent the molecular complexity of the compound. For examples, most of the very simple inorganic compounds belong to the regular and hexagonal systems. Orthorhombic and monoclinic systems are favored as the molecular structure becomes more complex, and 60% of the natural minerals belong to these systems (Mullin, 1997; Myerson, 1993).

The habit of a crystal grown in solution depends on solution conditions such as solvent, temperature and supersaturation, as well as the type and amount of impurities present. Impurities that modify crystal habit can be divided into the following categories: ions, ionic surfactants, nonionic surfactants like polymers and chemical bonding complexes. Ionic surfactants, for example, are routinely used in industry to control crystal morphology during crystallization. These impurities can compete with the crystallizing molecules for adsorption sites on the faces of the growing crystals. In the absence of impurities, the crystal morphology depends on the energy of attraction between the various crystal faces and the depositing molecules, i.e., the relative rates at which growth occurs at individual faces. If adsorption of impurities is selective, i.e., only to specific crystal faces, or to different extents at different faces, significant changes in the growth rates on those faces will to lead to habit changes (Long, 1995; Mullin, 1997). Such influence of impurities on calcium oxalate crystallization is further discussed in section 4.3.

2.3 Particulate Deposition

The deposition of particles from flowing suspensions onto adjourning surfaces is of primary importance in the fouling of heat exchange equipment and numerous other engineering applications such as membrane filtration systems. The rate and extent of particulate deposition are normally controlled by the following steps: particle transport from the bulk solution to the surface, attachment of particles on surface and particle re-entrainment (removal) from the surface (Epstein, 1999).

2.3.1 Particle Transport

The transport mass flux of particles from the mainstream to the deposition surface is expressed by:

$$W_t = K_t(c_b - c_s) \tag{2.32}$$

where K_t is the particle transfer coefficient, c_b and c_s are the concentrations of particles in the bulk and adjacent to the wall, respectively. The regimes of particle transport can be divided into diffusion, inertia and impaction based on the range of particle relaxation time, τ_p^+, which is a dimensionless measure of particle inertia and reflects how closely particles follow fluid streamlines (Epstein, 1999; Yiantsios and Karabelas, 1998):

$$\tau_p^+ = \frac{\rho_p d_p^2 u^{*2}}{18\mu\nu} \tag{2.33}$$

where ρ_p and ρ are the densities of particle and fluid, repsectively, d_p is the particle diameter, ν is the fluid kinematic viscosity, μ is the fluid dynamic viscosity and u^* (= $\bar{u}\sqrt{f/2}$ where $\bar{u}$ is the bulk fluid velocity and f is the Fanning friction factor) the wall shear velocity.

In the diffusion regime, $\tau_p^+ < 0.2$, colloidal particles (d_p < 0.1 μm) moving along the turbulent eddies in the main steam of the flow are carried to the wall by the Brownian motion through the viscous sublayer in the case of a turbulant flow. Thus the submicron particles can be considered as large molecules so that K_t is equivalent to the conventional mass transfer coefficient, K_m, which can be predicted from appropriate empirical or theorectical correlations for forced convection mass transfer such as those proposed by Webb (1971), Metzner & Friend (1958) and Epstein (1997,1999). In the inertia regime where $0.2 < \tau_p^+ < 20$ (d_p > 1-2 μm), the particles are large enough to obtain a transverse "free flight " velocity which is not completely dissipated in the boundary layer and carries the particles to the wall. Thus the transport of particles in this regime is by virtue of their inertia. As the particle size further increases, i.e., $\tau_p^+ > 20$ ($d_p \cong$ 10-20 μm), the particle transport enters the impaction regime and becomes relatively independent of the intensity of the turbulent eddies. In this regime, the particle velocity towards the wall approaches the friction velocity, u^*, and the particle stopping distance (S_d), the distance a particle with a certain momentum will travel in a quiescent viscous sublayer, becomes of the same order of the pipe diameter (Epstein, 1997; Papavergos and Hedley, 1984). Papavergos and Hedley (1984) developed an empirical corrleation for each of these regimes from expreimental data:

$$K_t^+ = 0.07Sc^{-2/3} \text{ at } \tau_p^+ < 0.2 \tag{2.34}$$

$$K_t^+ = 3.5x10^4(\tau_p^+)^2 \text{ at } 0.2 < \tau_p^+ < 20 \tag{2.35}$$

$$K_t^+ = 0.18 \text{ at } \tau_p^+ > 20 \tag{2.36}$$

where K_t^+ is the dimensionless particle transfer coefficient (= K_t/u^*) and Sc is the Schmidt number based on the Brownian diffusivity.

2.3.2 Particle Attachment

The attachment of particles at wall can be evaluated by two approaches. It may be statistically represented by the sticking probability, P, the probability that a particle transported to the wall sticks to it. Alternatively, a more fundamental approach can be considered by taking into account the interfacial forces involved between the particle and the surface (or deposit layer on surface), which include, but are not limited to, van der Waals forces, electrical double layer interaction, steric repulsion and hydrodynamic forces at wall (Epstein, 1999; Israelachvili, 1991; Turner, 1993).

2.3.2.1 Sticking Probability

Sticking probability is often introduced as a measure of adhesion force of a particle to the surface, assuming that adhesion occurs through chemical bonding between the particle and the surface. Thus P is postulated to have a typical Arrhenius dependence on the surface temperature. Other factors that may affect the magnitude of P include fluid velocity and particle size (Turner, 1993). Watkinson and Epstein (1970) proposed that the sticking probability be proportional to the ratio of adhesion force to wall shear stress, i.e.,

$$P \propto \frac{e^{-E/RT_s}}{\bar{\tau}_s} \tag{2.37}$$

where $\bar{\tau}_s$ is the average fluid shear stress at surface, T_s is the surface temperature and E is the activation energy for chemical bond formation. Beal (1978; 1983) analysed the particle deposition data from a variety of experiments. The correlations developed for all experimental conditions showed that the sticking probability remained constant for small particles but decreased with the dimensionless stopping distance, S_d^+, for relatively larger particles. For sand grain-water systems, the following relationships were derived:

$$P = 1,\ S_d^+ \le 2.4 \tag{2.38}$$

$$P = (2.4/S_d^+)^2,\ S_d^+ \ge 2.4 \tag{2.39}$$

2.3.2.2 Interfacial Forces

The two most important interfacial forces between colloidal particles or a colloidal particle and a surface are the London-van der Waals forces of attraction and the electrical double layer interaction forces. London-van der Waals forces arise largely from the fluctuating electromagnetic field produced from random charge density variations in the electron clouds of atoms. The resulting forces between particles and between a particle and a surface are always attractive in water. The double-layer interaction forces occur as a result of the overlap of the diffuse layers of charge that are developed adjacent to a charged surface, which together are called electrical double layer. The double layer forces arising from the overlap of the diffuse layer of charge can be repulsive or attractive depending on the sign of the charges of the interacting surfaces (Israelachvili, 1991).

In the case of significant repulsive double layer interaction (equal signs of a particle and surface), the total potential energy resulting from the combination of electrostatic and London-van der Waals forces will go through a maximum as the particle approaches the surface (Fig 2.3) according to the DLVO theory (developed by Derjaguin and Landau, and Verway and Overbeek to qantify the stability of colloidal suspensions). This barrier is analogous to the activation energy barrier for a chemical bonding reaction at the particle surface (eqn 2.37) and represents the energy that a particle must possess to overcome this barrier and deposit on surface (Epstein, 1997; Israelachvili, 1991). Some general properties of the total interaction energy with respect to particle deposition are as follows:

- If the energy barrier is not present (i.e., electrostatic attraction), particles will deposit at a transport limited rate due to van der Waals attraction.
- Particle deposition rates can be reduced by raising the energy barrier. This may be achieved by changes in the solution pH and the addition of specifically adsorbed species, which reverse the charge on the particles to the same sign as that on the surface.
- According to the classic DLVO theory, deposition at the primary minimum should be irreversible as the van der Waals attraction is infinitely strong at the point of contact between the surfaces (i.e., no particle removal). In practice, however, short-range effects (e.g., Born repulsions) are thought to limit the closest distance between particle and surface and consequently, the depth of the primary minimum (Oliveira et al., 1993).

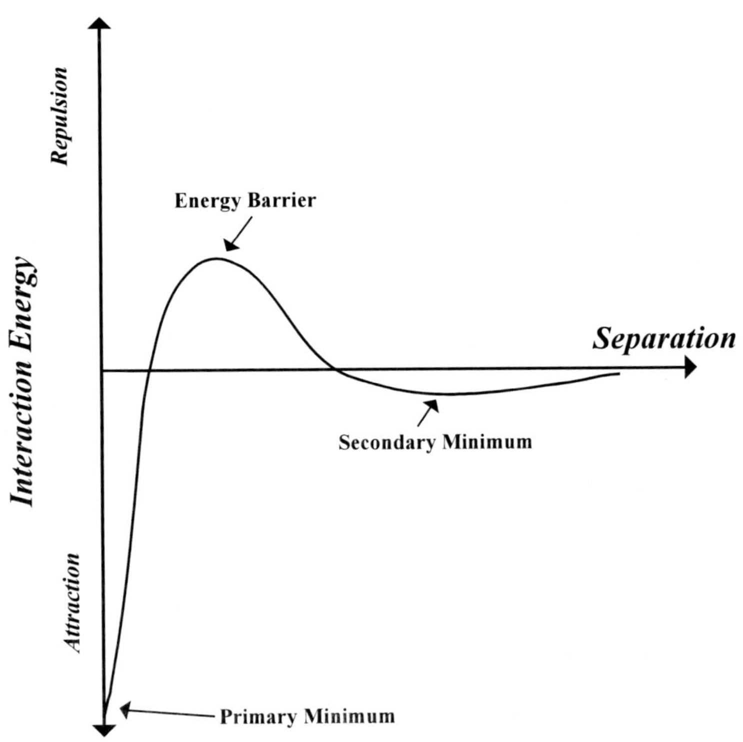

Fig. 2.3 Typical total energy profile when electrostatic repulsion is significant (arbitrary units)

- The secondary minimum, in which weak attachment is possible, can appear at large separation distances.

Steric stabilization is the stability imparted by polymeric molecules adsorbed onto or attached to a particle. As two polymer coated particles approach each other, a repulsive force is formed between the polymers as a result of mixing (osmotic) and volume restriction (elastic) effects. The former effect is related to unfavorable entropy due to increased density of polymer units from interpenetration of polymer chains between interacting surfaces, which restricts polymer-liquid interaction. The latter effect derives from the changes in the polymer configuration due to the compression of the adsorbed layers. Attached polymer molecules dangle out in solution and the confinement of these dangling chains upon approach of another surface results in a repulsive force. Steric interactions are dependent on the coverage and thickness of adsorbed polymer layer on each surface. If the adsorbed layer is thin, surfaces may be able to approach at close range where van der Waals forces can promote attachment (Gregory, 1989; Israelachvili, 1991).

It is generally accepted that hydrodynamic forces near the surface also play an important role in the deposition of particles on surface. These forces include tangential force (i.e., shear force), Saffman lift force (or near-wall lift force) and drainage force (or viscous interaction force). The magnitudes of these forces depend primarily on particle size (to the power of 2 or above) and the viscosity of media, and in most cases they tend to inhibit particle attachment and promote removal process. Detailed descriptions of these forces can be found in the references (Epstein, 1997; 1999).

2.3.3 Particle Removal (Re-Entrainment)

Contrary to the amount of work devoted to particle deposition to the wall, few researchers have attempted to speculate the mechanisms of particle removal from the surface. The most commonly used approach of modeling removal rate was based on the Kern and Seaton (1959) concept (see section 4.8) where the shear stress due to the velocity gradient within the boundary layer was considered to be the only important factor to cause particle re-entrainment. Another plausible explanation is that the re-entrainment occurs as a result of turbulent burst, which consists of random sequences of ejections from the wall boundary layer into the outer flow and sweeps of fluid from the outer region towards the wall (Kaftori et al., 1995). Cleaver and Yates (1975) suggested that the deposited particles could be re-entrained by the updraft generated during the fluid ejection. They derived their model based on statistical aspects of the bursts' occurrence in space and time by assuming that the bursts created sufficient lift to dislodge particles. Their results were in qualitative agreement with the limited experimental data. On the other hand, Yung et al. (1989) investigated the role of turbulent bursts in the re-entrainment process of particles from viscous layer using flow visualization. The results showed that a rolling mechanism due to horizontal drag force was more dominant than vertical lift forces, and turbulent burst activity was insignificant in particle entrainment, in contrast to the hypothesis of Cleaver and Yates. Reeks and Hall (2001) demonstrated the dominance of the drag force over the lift force by tangential and normal adhesive force measurements of alumina and graphite particles on a stainless-steel substrate subject to a fully developed turbulent flow. The drag force was found to typically exert a much greater effect by a factor of 100 over the lift force.

Furthermore, the structure of the deposit layer may undergo substantial changes at the surface which may or may not affect re-entrainment. For example, the stored thermal stress can induce plane of weakness in the crystalline deposits which encourages spalling (Bridgwater and Loo, 1984). Beal (1973) proposed a model that distinguished loose deposit removable by the circulating liquid from the hard interior deposit, which was not subjected to such removal. The author suggested that the loose deposit could be gradually compacted into hard deposit at the interface between the two. Quantitative details of this process, however, were not given.

Chapter 3

METHODOLOGY

3.1 BATCH CO-PRECIPITATION EXPERIMENTS

3.1.1 Materials

Stock solutions of calcium chloride, sodium oxalate and silica were prepared by dissolving known amounts of AR grade calcium chloride ($CaCl_2 \cdot 2H_2O$, Chem-Supply), sodium oxalate ($Na_2C_2O_4 \cdot H_2O$, Sigma) and sodium metasilicate ($Na_2SiO_3 \cdot 9H_2O$, Aldrich) in CO_2 - free distilled water, respectively. Synthetic juice solutions were prepared from a commercial sugar (as sucrose, purity 99.5%) kindly supplied by Sugar Australia, Pty, Ltd., NSW. All other reagents used were AR grade.

3.1.2 Methods

To match the conditions encountered in the latter effects (3rd – 5th) of a sugar mill evaporator to a certain extent, batch experiments were carried out at pH 6.0, 7.0 and 8.0 and at temperatures of 60°C, 70°C and 80°C. The initial concentrations for COM were between 20 ppm and 200 ppm and that of SiO_2 were between 50 ppm and 600 ppm. The initial molar ratio of calcium and oxalate ions was kept constant at 1.0, a condition that is known to favor the precipitation of COM (Ulmgren and Radestrom, 1997).

Spontaneous precipitation of COM from supersaturated pure solutions was conducted by mixing calcium chloride and sodium oxalate solutions without seeding (so the potential interfering effect of COM seeds on silica polymerization in the binary systems could be avoided.). The polymerization of SiO_2 in pure solutions was carried out by neutralising sodium metasilicate solutions to the required pH [sodium silicate is present in solution as monosilicic acid which dimerizes and then polymerizes to form silica at pH 6-8 (Sheikholeslami et al., 2001; Sheikholeslami and Tan, 1999)]. COM precipitation and SiO_2 polymerization reactions in sugar media were initiated by mixing supersaturated solutions of COM and SiO_2 with sugar solutions (10-40% (wt %) sugar in final mixtures, a range covering that of juice concentrations entering the 3rd – 5th effects of a quintuple evaporator set in a sugar mill). The initial pHs of individual solutions and mixtures were adjusted to desired value using dilute HCl (APS) or NaOH (Chem-Supply). All COM and SiO_2 solutions were then transferred to capped plastic tubes and placed in the water bath until equilibria were reached (or for at least 24 h).

Kinetic experiments of both COM and SiO_2 precipitation were conducted to examine the time needed to reach equilibrium. For COM precipitation, aliquots of sample solutions were withdrawn at predetermined intervals and immediately filtered through a syringe end membrane filter (0.22 μm Millipore Corporation). To avoid further reaction after the filtration, filtrates were treated with concentrated HCl to give a pH of ~1.5 (The solubility of COM significantly increases at pH <2) (Ulmgren and Radestrom, 1997) before analysis.

Similar sample preparation procedures were followed for the batch tests of SiO_2 polymerization except that the filtration step was omitted because it was difficult, if not impossible, to differentiate between dissolved and colloidal silica with 0.22 μm membrane filter (Millipore). Instead, different analytical methods were used to characterize SiO_2 in solution based on its reactivity (Sheikholeslami and Tan, 1999). The silica molybdate colorimetric method, for example, was used to selectively determine soluble silica species ('reactive' silica) because of its capability to form rapidly the silica molybdate complexes under acidic conditions. Inductively Coupled Plasma (ICP) Spectroscopy was used for the measurement of 'total' SiO_2 content, a collective term including all reactive (soluble) and non-reactive (colloidal) forms of SiO_2, assuming that all small colloidal SiO_2 particles were destroyed in the plasma and quantified along with soluble species. Thus, the difference in the concentration of SiO_2 measured by these two methods can be interpreted as the evidence for the appearance of colloidal SiO_2 particles in solution.

- In general, samples were analyzed at room temperature using the following procedures:
- pH was determined using an Orion 370 pH meter and combination pH electrode calibrated with standard buffers;
- Calcium concentrations and total silica (as Si) concentrations were determined using a Varian Vista AX Simultaneous Inductively Coupled Plasma Atomic Emission Spectrometer (ICP-AES) connected to a SPS-5 autosampler;
- Oxalate concentrations were determined by UV Visible spectrophotometer (Varian Cary 1E) using a colorimetric method (Burriel-Marti et al., 1953);
- Concentrations of reactive silica were determined by UV Visible spectrophotometer (Varian Cary 1E) using a silica molybdate complex colorimetric method (APHA, 1996).

Following the determination of thermodynamics and kinetics of COM precipitation and SiO_2 polymerization in single systems (i.e., water and sugar solutions), comparative studies on binary COM-SiO_2 systems were conducted by mixing neutralised sodium metasilicate solutions with freshly prepared supersaturated solutions of calcium oxalate in the presence or absence of sugar at initial pH 6.0, 7.0 and 8.0. Samples were equilibrated in water baths at 60°C, 70°C and 80°C and the concentrations of calcium, oxalate, reactive and total SiO_2 monitored with time.

The major experimental errors in determining the concentrations of COM and SiO_2 are composed of errors in dilution of the sample, approximation resulting from neglecting the influence of ionic strength on activity coefficients in diluted samples (see section 2.2.3), as well as errors associated with the calibration of the UV-visible spectrophotometer (UVV) and ICP-AES, and drift and fluctuation of their readings. Most of the experimental errors arose from UVV and ICP analyses (~6%) and ionic strength effect (~2.5%). Errors due to sample dilution were relatively small (<0.5%).

3.2 Dynamic Fouling and Boiling Heat Transfer Experiments

3.2.1 Test Apparatus

The experimental apparatus used for the dynamic runs is shown in Fig 3.1. It is comprised of a closed-loop circulation system which includes a storage tank, a circulation pump (Model CH4-20, Gundfos), a double-pipe heat exchanger, a cooling heat exchanger, series of pressure, temperature and flow measuring and control devices, optional micro-filters (20μm pore size, Cuno Pacific Pty. Ltd.) located before and after the annular test section for removal of crystals/particulates in the solution and a desk-top computer for control and data acquisition. It was designed for this work to operate under either constant composition mode to investigate the effects of operating conditions on the formation of composite scales, or under continuous evaporation to simulate the multi-effect evaporation cycle in a sugar mill.

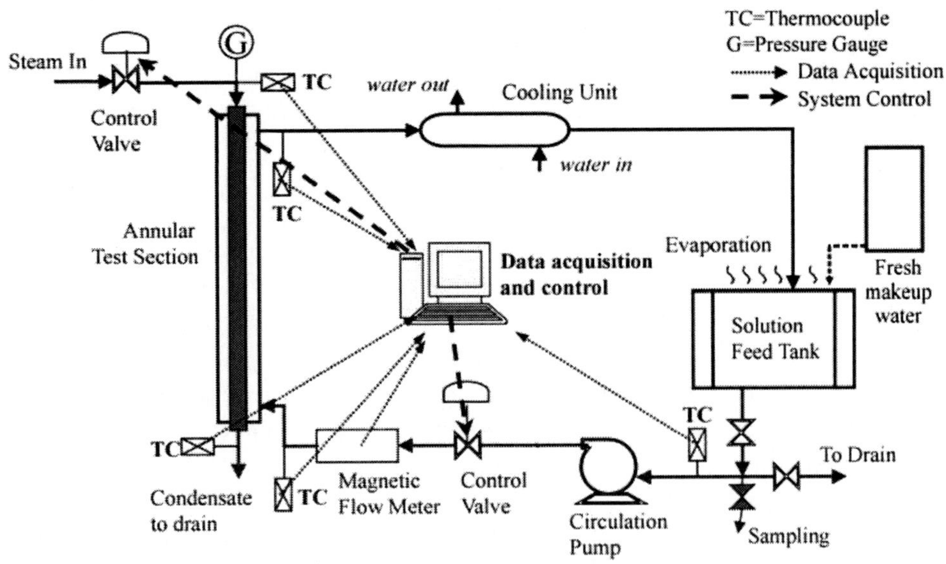

Fig. 3.1 A schematic diagram of test rig

The annular test section consists of a central heating tube made of stainless steel (Dimensions: 19.1mm outer diameter, 1.6m length, Sandvik Australia P/L)[1] and a glass outer wall (Inner diameter 38.1mm, Pegasus, Canada) to allow visual observation of the deposition

[1] Tests conducted by SRI found no marked difference in the nature of the scale formed inside a tube and those formed on the outside tube surface when heat is supplied in opposite directions. The sizes of calandria tube have no marked effect on the types of scale formed (Doherty 2004).

process during the experiment. All other pipings and fittings in the water systems were manufactured from stainless steel to prevent corrosion. The flow rate was measured by a magnetic flowmeter (Promag 50P, Endress & Hauser). The test solution was heated by counter-current flow of steam (100-200 kPa) obtained from a regulated steam supply system. An in line cooling unit (Diecon, Marine Products Inc.) was used to obtain the heat balance within the flow loop and keep the bulk temperature of the feed constant (60-80°C) throughout the run. The data acquisition and control system (Genie, American Advantech Co.) was used to collect inlet and outlet temperatures of both steam and water systems from embedded thermocouples and the water flow-rate from a magnetic flowmeter and also to control the water flow-rate and heat flux with PID regulators (Spirax Sarco P/L). The heat flux was maintained constant by controlling the temperature difference between water inlet and outlet (see eqn 3.3). The results were constantly displayed on the computer screen and recorded into the computer's hard disk at given time intervals. To reduce the magnitude of data fluctuation, all experimental measurements were averaged over every 10-20 min prior to further analysis.

3.2.2 Test Procedures and Conditions

3.2.2.1 General Procedure of Fouling Experiments

Dynamic tests were performed as described in the following. Distilled water or sugar solution in the feed tank (200-250L) was first heated by circulating through the annular test section and the system allowed to stabilize at predetermined operating conditions before known amounts of calcium, oxalate and silica (see section 3.1 for details of solution preparation procedures) pre-dissolved in portions of distilled water (2.5-10.0 L) were added successively with 5 minutes interval to allow complete mixing in the bulk solution and let the system re-stabilize. HCl or NaOH (0.1-10 M) was used to adjust the initial pH of the concentrated solutions to 6.5 before adding the chemicals to the feed tank. Data collection was then initiated after all the chemicals were mixed in the tank, this point being taken as t = 0. The changes in COM and SiO_2 concentrations during the run were monitored through periodic sample withdrawal from the test solution. The withdrawn samples were either immediately filtered through 0.22 μm Millipore syringe filters and acidified to pH <2 using concentrated HCl for later analysis of COM or were diluted without filtration for SiO_2 analysis. Calcium, oxalate, total SiO_2 and reactive SiO_2 contents were then determined using ICP-AES and UV-visible spectroscopy. Details of the sample preparation and analysis have been described in section 3.1.2. After each experiment, the scale was removed from the tube and sent for instrumental analysis (see section 3.3). The bulk densities of scales were estimated by weighing the different samples and measuring their volumes. The thermal conductivity of scales was estimated from the calculated fouling resistance and measurements of deposit layer thickness. New heat exchanger tubes were used for each experiment. The tube surfaces were cleaned with 2% (w/v) hot caustic soda followed by dilute acid (pH =1) to remove dirt and grease prior to being installed in the loop.

3.2.2.2 Flow Boiling Heat Transfer Experiments

Experiments were conducted to evaluate the nucleate flow boiling heat transfer coefficients in water and sugar solutions. Table 3.1 shows the range of variables used in the experiments as compared to those of the sugar mill evaporators. The current fouling loop system was limited to flow boiling runs with higher subcooling levels than those typical of latter effects of sugar mill evaporators because of limited temperature range of equipments. Prior to flow boiling experiments, the fluid wettabilities were analyzed by taking photographs of single droplets (>10) of water and sugar solutions on the heat exchange tube using a CCD camera and measuring the static contact angle of each fluid.

Table 3.1 The ranges of operating parameters used in flow boiling heat transfer experiments in comparison to those in the sugar mill evaporator (Doherty, 2002; Lavarack, 2001)

Condition	This work		Sugar mill evaporator (1st – 5th effect)
	Water	Sugar solution	
Heat flux (kW/m^2)	45.6-112.7*	50.7-86.9*	16.7-23.3
Velocity (m/s)	1.0-1.8	0.8-1.6	0.8-2.8
Subcooling (K) $^{\alpha}$	12.0-35.0$^{\#}$	16.0-35.0$^{\#}$	0-17.0
Superheat (K) $^{\beta}$	4.6-17.4	4.1-24.7	12.0-26.0
Sugar conc. (wt %)	0	25-50	15-68

$^{\alpha}$ difference between the localized saturation and the localized bulk temperatures of the fluid
$^{\beta}$ difference between the wall and the localized saturation temperatures of fluid
* higher ranges used due to flow boiling without vacuum system
$^{\#}$ higher subcooling values used due to limited temperature range of equipments

3.2.2.3 Evaporative Concentration Fouling Experiments

To simulate various effects of the evaporation cycle in a sugar mill evaporator, the test solution was concentrated by evaporation (~5-6 L/h) without passing through either of the in line filters during the run and the key operating conditions such as superheat, COM/SiO_2 supersaturation (SS) and concentration factor (CF, i.e., the degree of concentration) maintained in the range comparable to those usually encountered in the later effects of a sugar mill evaporator. Single and binary solutions were tested using identical experimental conditions. In the binary systems, COM was used as the controlling species, the initial supersaturation of which was varied between 1-5.3 to determine the maximum degree of composite fouling from various feed solutions, in the presence of a fixed initial supersaturation of SiO_2 (1.7) (Appendix IV). Table 3.2 presents a summary of the operational parameters of later effects of a sugar mill evaporator in comparison to the experimental conditions of this work. The continuous evaporation process was used to determine which stage of evaporator resulted in the onset of COM/SiO_2 fouling and the corresponding critical SS values. In addition, this will enable further investigation on the effect of other factors such as surface temperature and fluid velocity to be carried out under a constant solution composition, as described in the next section.

Table 3.2 Operational conditions of evaporators in a sugar mill (Doherty, 2002; Honig, 1963) in comparison to this work

Conditions	Sugar mill evaporator 3rd effect	4th effect	5th effect	This work
Conc. factor (CF)	1.5*	2.0*	3.0*	1-10
Initial (feed) supersaturation				
COM [a]	2-7	4-12	5-22	1-5.3
Silica [b]	1-5	2-6	3-11	1.7
Overall range of supersaturation (3rd effect feed – 5th effect exit)				
COM	2-66			1-53
Silica	1-33			1-17
Initial COM/SiO_2 SS ratio (3rd effect feed)	0.5-3.5			0.6-3.0
Temperature (°C)	100	80	54	80
Pressure (kPa)	98	50	15	101
Superheat (°C)	12	20	26	10

* Approximate values based on the changes in the sugar conc. (wt %) in a sugar mill evaporator.
[a] COM supersaturation is calculated as: [COM]/19 (19 is the solubility (ppm) of COM at 80°C).
[b] Silica supersaturation is calculated as: [SiO_2]/290 (290 is the solubility (ppm) of SiO_2 at 80°C)

3.2.2.4 Constant Composition Fouling Experiments

Additional experiments were performed under a constant solution composition and nucleate flow boiling heat transfer to evaluate the effects of thermal hydraulic parameters on composite fouling (Appendix V). Table 3.3 shows the summary of operating conditions used in this work in comparison to those in the sugar mill evaporators. A fixed sugar concentration of 25% (wt %), which is typical of the juice concentration exiting the No. 2 effect of a quintuple evaporator set in sugar mill, was used in all experiments. The initial concentrations of COM and SiO_2 were determined based on the results of previous runs of evaporative concentration, which suggested that composite fouling at an initial COM concentration of 50 ppm (initial SiO_2 concentration 500 ppm) could start as early as when the concentration factor reached ~1.7 (see section 4.5). Typical duration of each run was ~150-160 hrs, beyond which significant sucrose degradation may occur as was evidenced by color formation (Eggleston and Vercellotti, 2000), and was adequate for the present purposes. The water loss due to evaporation in the main tank was compensated by a feed of fresh makeup water regulated by a level control system, which includes a proportioning solenoid valve and pressure transmitter (Burkert Contromatic Australia P/L). Fresh chemicals were also added to the main tank at an average rate of 1.9 ppm COM /h and 2.7 ppm SiO_2 /h to maintain the concentrations of COM and SiO_2 in solution, which otherwise tended to decrease with time due to deposit buildup on the heat exchange surface and bulk precipitation.

Table 3.3 The ranges of operating parameters used in constant composition runs in comparison to those in the sugar mill evaporator (Doherty, 2002; Lavarack, 2001)

Condition	This work	Sugar mill evaporator (1st – 5th effect)
Heat flux (kW/m^2)	56.4-104.9	16.7-23.3
Superheat (K)	5.0-15.0	12.0-26.0
Velocity (m/s)	0.8-1.8	0.8-2.8
Reynolds number (liquid phase)	7875-18121	n/a
Subcooling (K)	20.0-30.0	0-17.0
Bulk temperature (°C)	70.0-80.0	54.0-100.0

3.2.3 Data Treatment

The instantaneous fouling resistance, $R_{f(t)}$ (m^2 K/kW) is calculated (eqn 3.1) from the overall heat transfer coefficient (U_o) at the beginning of the each run and at any given time, t, when fouling has taken place. The overall heat transfer coefficient, U_o (W/m^2 K), is determined from the information on the heat duty, q (W), the heat transfer surface area, A (m^2) (based on outside diameter of the heat transfer tube) and ΔT_m, the log-mean-temperature-difference (K) for a counter current arrangement (eqn 3.2). The assumptions made in using eqn (3.2) include: steady-state flow of heat, constant overall heat transfer coefficient over the range of temperature tested, and no heat loss to the surrounding. The use of ΔT_m term further implies that any axial dispersion of heat has been neglected.

$$R_{f(t)} = \frac{1}{U_{o(t)}} - \frac{1}{U_{o(t=0)}} \tag{3.1}$$

$$U_o = \frac{q}{A\Delta T_m} \tag{3.2}$$

The heat transfer rate can be expressed in terms of a fluid temperature difference by applying an enthalpy balance:

$$q = \dot{m} c_p \Delta T_l \tag{3.3}$$

where $\dot{m}$ is the mass flow rate (kg/s), c_p is the specific heat of heating fluid (J/kg °C) and ΔT_l is the temperature difference of fluid along the annulus (K).

To compare the effects of thermal hydraulic operating conditions, the instantaneous net fouling rate, ϕ_f (m^2 K/kW·s), was calculated by:

$$\phi_f = \phi_d - \phi_r = \frac{dR_{f(t)}}{dt} \tag{3.4}$$

where ϕ_d (m^2 K/kW·s) and ϕ_r (m^2 K/kW·s) are the deposition and removal rates, respectively. Depending on the relative extents of deposition and removal rates, the $R_{f(t)} \sim t$ fouling curve can show linear, falling rate or asymptotic behaviour. For particulate fouling, the deposition term is usually considered to be constant while the removal term increases with the deposit growth resulting in an asymptotic-type fouling curve (Kern and Seaton, 1959; Oliveira et al., 1993). The linear fouling kinetics generally indicates hard and adherent deposits due to crystallization fouling, while the falling rate may result from a falling deposition rate and/or increasing removal rate. It has to be noted, however, that the linear or falling rate fouling kinetics of a species may actually be the early stage of asymptotic behaviour, and sufficiently long period of fouling experiment would be required to be confident of its real fouling behaviour (O'Callaghan, 1986).

The subcooled flow boiling heat transfer (water-side) film coefficient, h, was obtained based on the resistance equation including the system geometry (Kreith, 1973):

$$\frac{1}{U_o} = \frac{1}{h} + \frac{\delta_w D_o}{k_w \overline{D_L}} + \frac{D_o}{D_i h_{st}} \quad . \tag{3.5}$$

where k_w and δ_w are the wall thermal conductivity (W/m K) and thickness (m), respectively. D_i is the inside wall diameter, D_o is the outside wall diameter and $\overline{D_L}$ is the log mean diameter. The steam-side heat transfer coefficient, h_{st}, was calculated assuming film-type condensation of steam (see Appendix IIIB for details).

The measurement uncertainty of temperatures was ±0.3°C and for velocity measurements this was about 0.5%. The uncertainties involved in heat flux and heat transfer coefficients were less than 9%, which was a typical range for this type of experiment (see section 5.2.2.1 for reproducibility data).

3.3 Instrumental Analysis of Deposit

3.3.1 Preparation of Deposits

Upon completion of the experiments, the deposits were either separated from the solutions by vacuum filtration through a Millipore membrane filter (0.22μm) (for batch tests) or scraped off the heat transfer surfaces (for dynamic tests), washed with distilled water before being filtered again under vacuum, dried to constant weight under room temperature and kept in a desiccator for further analyses. The density of scales formed in dynamic tests was estimated by weighing the different samples and measuring their volumes (using a graduated cylinder). The thermal conductivity of scales was estimated from the calculated fouling resistance and measurements of deposit layer thickness.

3.3.2 XRD Analysis

The composition and crystalline phases of the precipitates were analysed by XRD over the range 3 to 70 deg 2θ using a Siemens D5000 diffractometer with CuK_a radiation (40 kV and 30 mA) at a scanning speed of 1.0 deg 2θ/min and a step size of 0.02 deg 2θ.

3.3.3 SEM-EDS Analysis

A Hitachi 4500II field emission scanning electron microscope (SEM) in conjunction with an Oxford Instruments Link ISIS 200 energy dispersive X-ray spectrometer (EDS) was used for micrographic imaging and elemental analysis of precipitates. Samples were first dispersed in distilled water on a double-sided carbon tape attached to the aluminium stub. After water evaporation, the dry crystals were coated with copper to obtain a 200 Å thickness before SEM evaluation. For quantitative determination of elements present in the precipitates, samples were either treated with resin setting (for particle cross-section) or made into KBr pellets (for particle surface) and subsequently copper-coated in order to produce a flat and conductive surface required for EDS analysis. The resin setting was carried out at 20°C by embedding sample particles into plastic molds filled with fresh Spurr resin for polymerisation. Particle cross-sections were exposed using a Reichert-Jung Ultracut ultramicrotome with diamond knives.

The following conditions were used in the SEM-EDS analyses:

- Accelerating voltage of 20 keV for both SEM imaging and EDS analysis
- Working distance of 20.0 mm for EDS detector (focal point)
- Acquisition time at 50s (live time)
- Magnifications between x1000 to x20000

3.3.4 Fourier Transform Infrared (FT-IR) Analysis

Infrared spectra of CaOx and CaOx-SiO_2 composite were recorded on a Nicolet Avatar 320 FT-IR spectrophotometer (4000 - 400 cm^{-1} range; 4 cm^{-1} resolution) to obtain structural and bonding information on CaOx molecules. Disks of 100 mg of KBr containing ~1% of finely ground powder of each sample were prepared before analysis.

3.3.5 Zeta-Potential Analysis

The zeta-potential values of scales obtained from dynamic tests were determined from electrophoretic mobility measurements according to the Smoluchowski equation using a Brookhaven ZetaPALS instrument. Samples were prepared by adding scale particles into appropriate amounts of distilled water or sugar solution and allow the mixtures to settle for at least 4 hours before use (to ensure that particles in solution were within colloidal size range). The samples were then transferred to a glass cuvette (total sample volume ~1.5 mL), in which the electrode assembly was inserted and the mobility measurements made (electric field between 6.6-10.5 V/cm) at 25°C.

3.3.6 Particle Size Distribution Analysis

In order to further verify the existence of colloidal species in solution in addition to ICP/UVV analysis (see section 3.1.2/p36), the particle size distribution of COM-SiO_2 complexes formed in sugar solution was determined by passing the solutions sequentially through a series of filters of decreasing pore sizes: 2.5 μm filter paper (Whatman No. 41), 0.22, 0.05 and 0.005 μm membrane filters (Millipore) and analysing the concentrations of COM and SiO_2 in each filtrate. To prevent further reactions after each filtration, the temperature of the receiving vessel was held the same as that of the water bath. Small amounts of each filtrate solution were withdrawn and immediately diluted for SiO_2 analysis, or transferred to 10ml ICP tubes, in which one drop of concentrated HCl is added in advance, for determination of COM content.

Chapter 4

CO-PRECIPITATION OF CAOX-SIO$_2$ IN WATER AND SUGAR MEDIA

4.1 COM-SIO$_2$ CO-PRECIPITATION IN AQUEOUS SOLUTIONS

Although the deposition of SiO_2 and COM on the calandria tubes of evaporators causes serious processing problems in many cane sugar mills, previous studies on scale formation have concentrated on fouling of individual compounds and the development of inhibiting methods for each component. There is no information in the literature on the mechanisms of composite fouling of SiO_2 and COM. This section aims to evaluate the interactions between COM and SiO_2 and how these interactions would influence the thermodynamics and kinetics of COM and SiO_2 precipitation in binary foulant systems. The results of preliminary batch tests are presented for the precipitation kinetics and thermodynamics of COM and SiO_2 in both single and binary systems in aqueous solutions.

4.1.1 COM Precipitation and SiO$_2$ Polymerization in Single Systems

4.1.1.1 COM Precipitation Kinetics

The rate equation generally used to describe the kinetics of COM precipitation can be expressed as:

$$-\frac{dC}{dt} = k_r \alpha^n \qquad (4.1.1)$$

where C is the molar concentration of COM, k_r is a rate constant, which is primarily a function of temperature and activation energy (see eqn 4.1.5), α is the relative supersaturation, and n is the order of reaction representing the mechanisms by which crystallization proceeds. Values of n near unity suggest a diffusion-controlled mechanism during crystal growth, whereas n close to 2 represents a surface-controlled mechanism. Polynuclear mechanisms occur at higher order ($n>2$) (Nancollas et al., 1991).

Relative supersaturation, α, is defined as (Nancollas et al., 1991):

$$\alpha = [(q_t^{1/2} - q_e^{1/2}) / q_e^{1/2}] \tag{4.1.2}$$

where q_t and q_e are the ionic products ($(mol/L)^2$) of calcium oxalate at time t and at equilibrium respectively). After combining eqn 4.1.2 with eqn 4.1.1 and applying boundary conditions, $t = 0$ to $t = t$ and $q_t = q_0$ to $q_t = q_t$, the integrated form of eqn. 4.1.1 becomes:

$$q_t = \frac{q_0 - q_e}{\left[1 + [(q_0 - q_e)/q_e]^{n-1} \dfrac{1-n}{q_e} k_r t\right]^{1/n-1}} + q_e \tag{4.1.3}$$

This equation has been used to calculate the kinetic parameters of COM precipitation at constant pH and temperature by assuming the formation of only one CaOx crystal phase (COM) in all single and multi-component solutions tested (sections 4.1 and 4.2).

Figure 4.1.1 shows the precipitation kinetics of COM at 60°C. Under the current experimental conditions, the observed spontaneous precipitation of calcium oxalate monohydrate from a supersaturated solution was rapid and the equilibrium was reached within several hours of reaction.

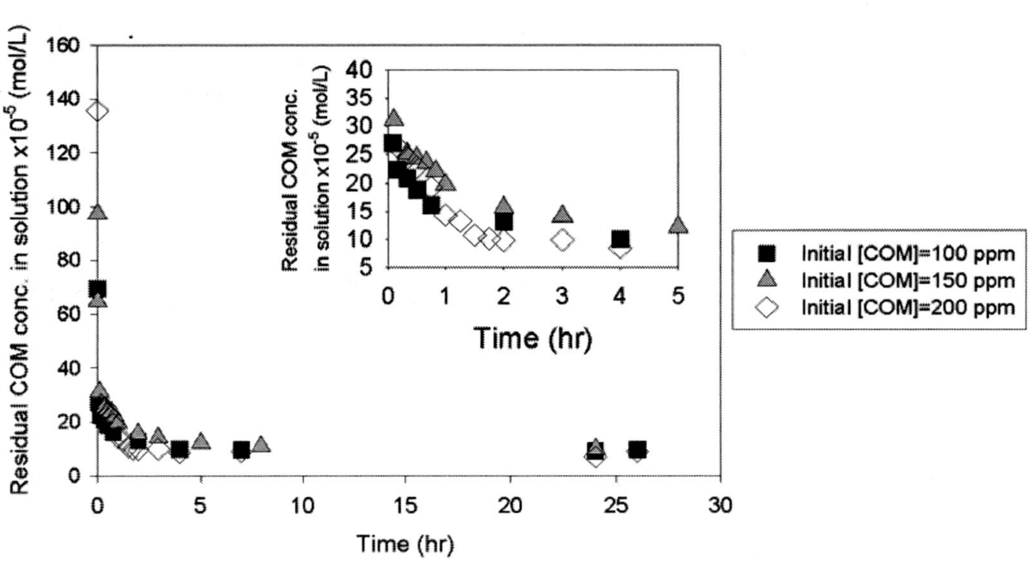

Fig. 4.1.1 The kinetics of COM precipitation at different initial COM concentrations at pH 8.0 and at 60°C.

The kinetics of COM precipitation at various initial COM concentrations (Fig. 4.1.1) were analyzed by fitting the rate equation (eqn. 4.1.3) to experimental data in order to calculate the values of the rate constant (k_r) and the order of reaction (n). These data are summarized in Table 4.1.1. The results show that an increase in the initial COM concentration at 60°C led to an increase in k_r and a corresponding decrease in n due to increased thermodynamic driving force. For a given initial COM concentration (150 ppm), the n values were found to decrease with increasing temperature while the respective rate constants increased. The n values were found to be in the range of 2.4-4.2, which were similar to those previously reported by Nielson (1960) but higher than those obtained by other workers (Garside et al., 1982; Nancollas, 1983). The difference may be accounted for, at least in part, by the differences in the experimental conditions (e.g., initial supersaturation ratio, seeding and mixing conditions, etc.) used in these studies.

Table 4.1.1 Calculated parameters for COM precipitation kinetics at different initial COM concentrations and temperatures and at pH 8.0 (Appendix IA and B)

System designation	Log (k_r)	n	Correlation co-efficient (R^2)
60°C			
Initial [COM]: 100 ppm	-0.12	4.2	0.996
Initial [COM]: 150 ppm	0.51	3.4	0.990
Initial [COM]: 200 ppm	0.63	3.0	0.995
70°C			
Initial [COM]: 150 ppm	0.69	2.5	0.998
80°C			
Initial [COM]: 150 ppm	0.85	2.4	0.999

The variation of the rate constant as a function of temperature can be described by the Arrhenius equation (Vyazovkin, 2000):

$$\frac{d(\ln k_r)}{dT} = \frac{E}{RT^2} \tag{4.1.4}$$

or in an integrated form

$$k_r = Ae^{-E/RT} \tag{4.1.5}$$

where E is the activation energy (kJ/mol), R is the gas constant (J/K mol), A is the "apparent" pre-exponential factor based on unseeded conditions and T is the temperature (K).

Taking the logarithms of eqn. 4.1.5 and incorporating the value of R (8.314 J/K mol) gives

$$\log k_r = \log A - \frac{E}{2.303RT} = \log A - \frac{E}{19.14T} \tag{4.1.6}$$

from which the activation energy can be readily calculated by plotting log k_r against 1/T to obtain a straight line with a slope of -E/19.14. The E value so calculated was *ca.* 38 ± 4 kJ/mol for COM. The activation energy could be used for mechanistic interpretation of crystallization kinetics. It has been reported that E values of ~12 and ~40 kJ/mol corresponded to bulk controlled and surface controlled processes, respectively (Nancollas, 1983). Thus, the activation energy obtained in this work demonstrates that COM crystallization does undergo a surface controlled process. This supports results in the literature (Garside et al., 1982).

4.1.1.2 COM Precipitation Thermodynamics

The observed solubility of calcium oxalate in dilute aqueous solution, expressed in terms of concentration solubility product, K_{csp} ($(mol/L)^2$) is given as:

$$K_{csp} = [Ca^{2+}]_{eq}[Ox^{2-}]_{eq} \quad (4.1.7)$$

where $[Ca^{2+}]_{eq}$ and $[Ox^{2-}]_{eq}$ are the measured concentrations of calcium and oxalate ions (mol/L) at equilibrium, respectively. The solubility concentration of COM, c*, therefore, can be written as:

$$c^* = \sqrt{K_{csp}} \quad (4.1.8)$$

Figure 4.1.2 shows the changes in the observed COM solubility concentrations obtained from eqns. 4.1.7 and 4.1.8 at different initial COM concentration and initial pH levels. As the initial COM concentration increased, the solubility concentration of COM exhibited a maximum at an initial COM concentration of ~50 ppm and then levelled off at ~90-100 ppm (Fig 4.1.2). This increase is probably due to the formation of calcium oxalate trihydrate (COT), which has a higher solubility than COM (Ulmgren and Radestrom, 1997). It might also be explained by the formation and incomplete precipitation of COM particles of colloidal sizes (particle sizes < 0.2 μm, which were not removed by the filtration process and were subsequently analysed by ICP) at low COM supersaturation levels. However, the latter assumption needs to be examined through further experiments.

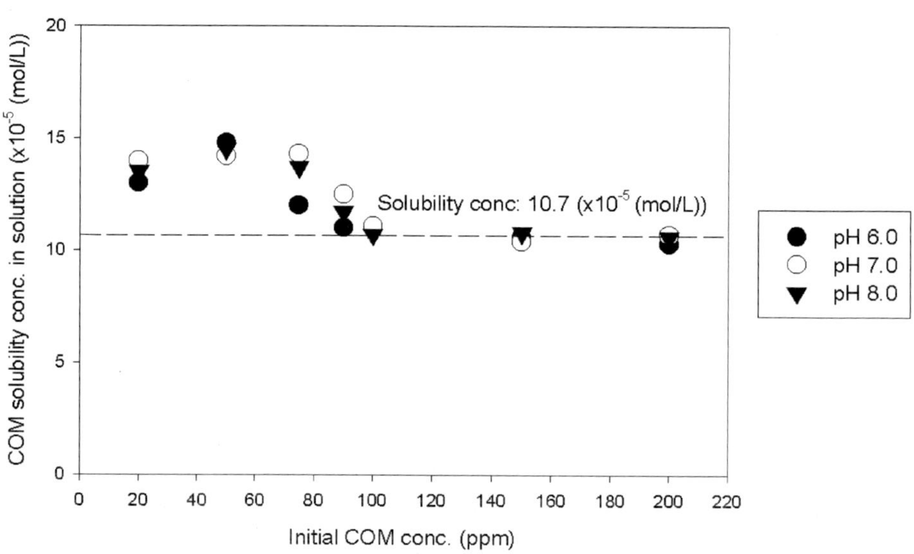

Fig. 4.1.2 The solubility concentration of COM as a function of initial COM conc. at different pH and at 70°C.

The derived K_{csp} and c* values for COM at different levels of pH and temperature (see Appendix IC-IE), as well as the reported literature values are given in Table 4.1.2. The solubility concentrations were not found to be dependent on pH but increased with increasing temperature between 60 – 80°C. The differences between the solubility concentrations obtained in this study and those reported in the literature could be due to, as mentioned before, the variations in the experimental conditions and/or different methodologies used for sample preparation and analysis.

The effect of temperature on the precipitation equilibria of salts can be expressed by the van't Hoff equation. This equation relates the value of the thermodynamic equilibrium constant, K_{sp}, to temperature as a function of the enthalpy of reaction (heat of precipitation) (Atkins, 1994):

$$\frac{d(LnK_{sp})}{d(1/T)} = \frac{-\Delta H^{\Phi}}{R} \tag{4.1.9}$$

where $K_{sp} = K_{csp} f_2^2$, where f_2 is the activity coefficient of a divalent (calcium or oxalate) ion. ΔH^{Φ} is the heat of precipitation (kJ/mol), R is the gas constant and T is the temperature (K). Assuming that ΔH^{Φ} remains constant within the range of temperature tested, a straight line is obtained if –LnK_{sp} is plotted against 1/T. The calculated ΔH^{Φ} in the 60-80°C range for COM using eqn 4.1.9 was found to be ~–29.8 kJ/mol, which was higher than –22.9 kJ/mol obtained by Nancollas and Gardner (1974). This could probably be explained by a lower temperature range (15-45°C), among other experimental conditions, on which Nancollas and co-worker's result was based, since ΔH^{Φ} is expected to increase at higher temperature ranges (Atkins, 1994) [e.g., ~65% increase in ΔH^{Φ} in the range 25-80°C for COM (Velich et al., 1992)].

Table 4.1.2 The solubility concentration, c* and K_{csp} values for COM in comparison with literature data

	Temperature (°C)					
	60		70		80	
	K_{csp} x10^{-9}(mol/L)^2	c* x10^{-5}(mol/L)	K_{csp} x10^{-9}(mol/L)^2	c* x10^{-5}(mol/L)	K_{csp} x10^{-9}(mol/L)^2	c* x10^{-5}(mol/L)
pH 6	9.3	9.6	11.0	10.5	17.0	13.0
pH 7	9.2	9.6	11.6	10.8	17.0	13.0
pH 8	8.7	9.3	11.5	10.7	16.4	12.8
Average	9.1	9.5	11.4	10.7	16.8	12.9
Literature data	--	8.8 [a]	--	10.1[a]/12.0[b]	--	11.5 [a]

[a](Stephen and Stephen, 1963)
[b](Ulmgren and Radestrom, 1999a)

4.1.1.3 Silica Polymerization Kinetics

The rate equation commonly used to describe the changes in reactive SiO_2 concentration during SiO_2 polymerization can be expressed as:

$$-\frac{d(SiO_2)}{dt} = k_r[(C_t - C_e)/C_e]^n \quad (4.1.10)$$

where C_t and C_e are the reactive silica concentrations (ppm) at time t and at equilibrium, respectively. k_r and n are the rate constant and order of reaction, respectively. The values obtained (Table 4.1.3) for the order of reaction using eqn 4.1.10 are mainly in the range of 1-2, which correspond well with data given in the literature (Chan, 1989).

Table 4.1.3 Calculated parameters * for SiO_2 polymerization kinetics for different initial SiO_2 concentration, pH and temperature levels (Appendix IF-IH)

Temperature (°C)	System designation	Log (k_r)	n	Correlation co-efficient (R^2)
60	Initial pH 8.0			
	Initial [SiO_2]: 350 ppm	-9.0	2.7	0.924
	Initial [SiO_2]: 400 ppm	-2.1	1.3	0.995
	Initial [SiO_2]: 450 ppm	-1.1	1.3	0.993
	Initial [SiO_2]: 500 ppm	-0.4	1.3	0.974
	Initial pH 7.0			
	Initial [SiO_2]: 400 ppm	-4.3	1.9	0.991
	Initial pH 6.0			
	Initial [SiO_2]: 400 ppm	-10.9	3.1	0.880 **
70	Initial pH 6.0			
	Initial [SiO_2]: 470 ppm	-3.9	1.8	0.930
80	Initial pH 6.0			
	Initial [SiO_2]: 520 ppm	-1.8	1.6	0.945

* kinetic parameters are calculated by data fitting of an integrated form of eqn (4.1.10) similar to eqn (4.1.3) ** poor data fit due to long induction period in data (Fig 4.1.4)

Figures 4.1.3 and 4.1.4 illustrate the dependence of silica polymerization kinetics on the initial silica concentration and pH at 60°C, respectively. The polymerization equilibria were attained after 14 days and a general trend of increasing polymerization rate with increasing initial silica concentration an d pH was observed within the range tested. Similar results have been reported by other workers (Rothbaum and Rohde, 1979; Sheikholeslami et al., 2001). At pH 8.0, the increment of polymerization rate was less obvious between 400-500 ppm, where the order of reaction displayed a plateau of 1.3 as the rate constant slightly increased. The dotted line in Fig. 4.1.4 marks the value of solubility of amorphous silica (220-230 ppm at 60°C) reported in the literature (Rothbaum and Rohde, 1979), showing good agreement with the present work. Figures 4.1.5 and 4.1.6 show that SiO_2 polymerization at a constant initial SiO_2 supersaturation of 1.8 (the initial SiO_2 concentration varied due to the changes in the SiO_2 solubility with temperature, see next section) was accelerated by an increase in the temperature from 60 to 80°C, again consistent with literature data (Rothbaum and Rohde, 1979).

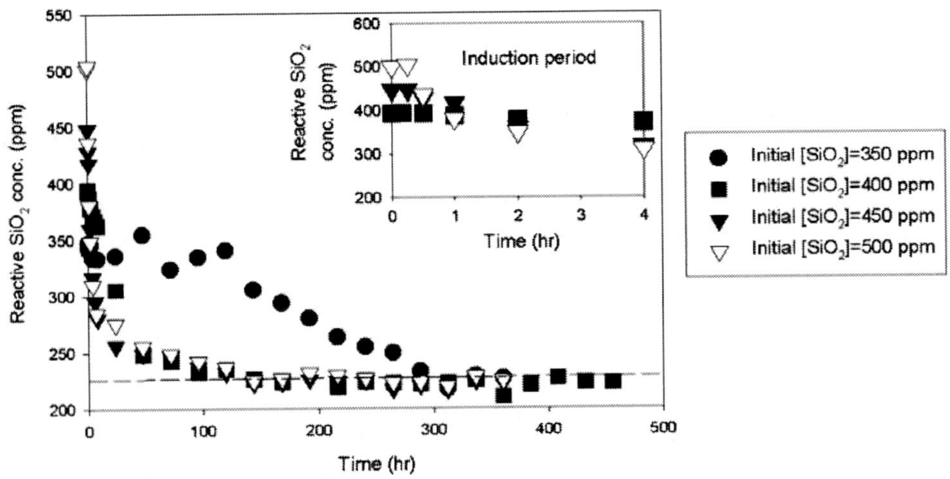

Fig. 4.1.3 The kinetics of SiO$_2$ polymerization at different initial SiO$_2$ conc. at initial pH 8.0 and at 60°C.

The polymerization of amorphous silica showed induction periods (a period when supersaturated silica solutions remain temporarily meta-stable without significant drops in dissolved silica concentrations) of various lengths, which were found to decrease with increasing pH (6.0-8.0), the degree of supersaturation (Figs 4.1.3 and 4.1.4, inset) and temperature (Fig 4.1.6). It has been suggested that the lengths of induction periods were sensitive to changes in pH (Rothbaum and Rohde, 1979), as shown by data in the present study.

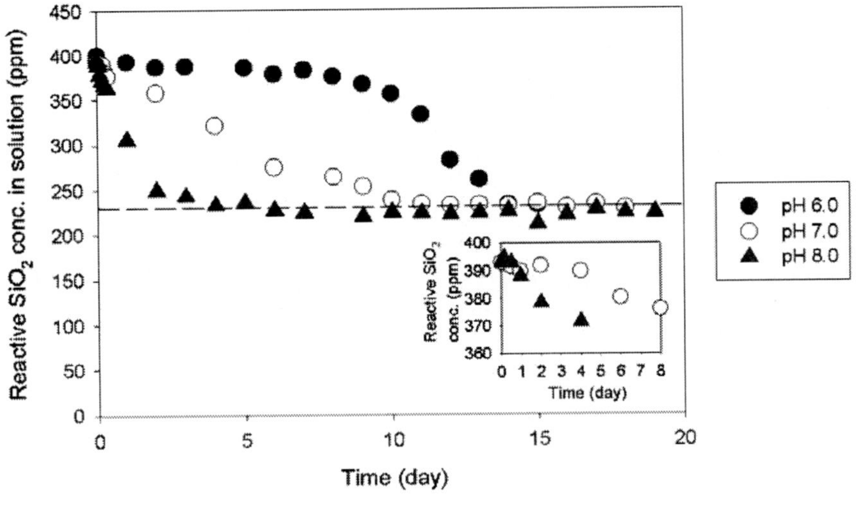

Fig. 4.1.4 Effect of initial pH on the kinetics of the polymerization of amorphous SiO$_2$ at initial SiO$_2$ conc. 400 ppm and at 60°C

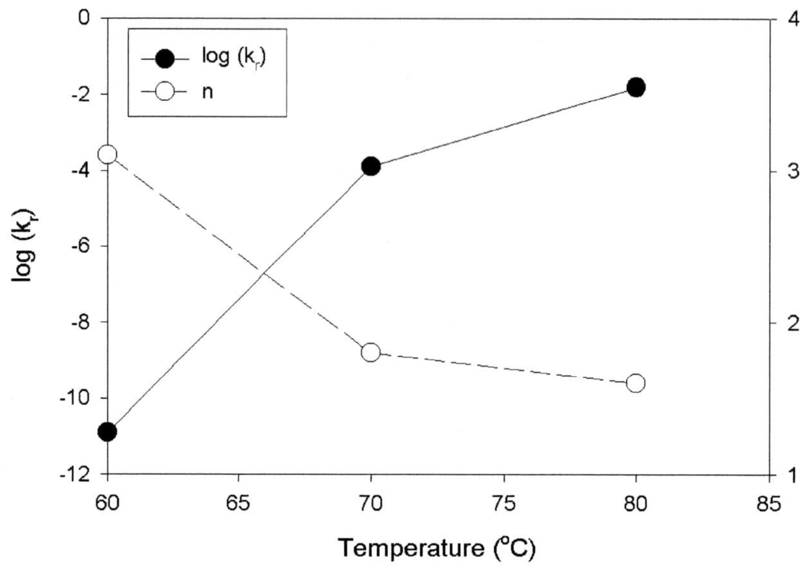

Fig. 4.1.5 Effect of temperature on the polymerization of SiO_2 at initial SiO_2 supersaturation 1.8 and pH 6.0

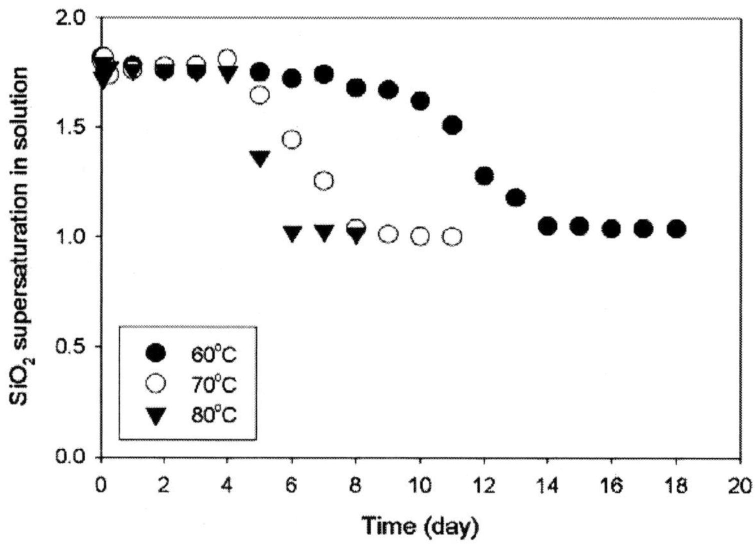

Fig. 4.1.6 SiO_2 polymerization kinetics at different temperatures, initial SiO_2 supersaturation 1.8, initial pH 6.0

4.1.1.4 Silica Polymerization Thermodynamics

Amorphous silica has been shown to deposit onto the surface in two different modes, depending on the level of initial supersaturation. There are:

- Direct or molecular deposition onto solid surfaces, and

- Homogeneous formation and growth of a colloid in the bulk or boundary layer and its subsequent deposition on the surface.

The latter process usually requires a supersaturation factor of about 2.5 or higher to take place (Chan, 1988; Sheikholeslami and Tan, 1999).

Figure 4.1.7 shows the effect of initial supersaturation on the type of silica deposition at 60°C (Appendix II). At high supersaturations (initial [SiO_2]: 450-600 ppm) data rose above the 1:1 line indicating the presence of non-reactive (e.g., colloidal) forms of silica in the solution, some of which eventually deposited leading to drops in the total SiO_2 concentrations. On the other hand, the direct deposition of dissolved silica onto tube surface was the only mechanism observed at lower supersaturation levels (initial [SiO_2]: 350-400 ppm). The formation of SiO_2 colloids and bulk deposition was able to occur from an initial supersaturation ratio of 2.0 (initial SiO_2 concentration: 450 ppm) instead of 2.5 as was found by Chan (1988), probably because our experiments were conducted at higher temperature, which resulted in high reaction rate (Rothbaum and Rohde, 1979).

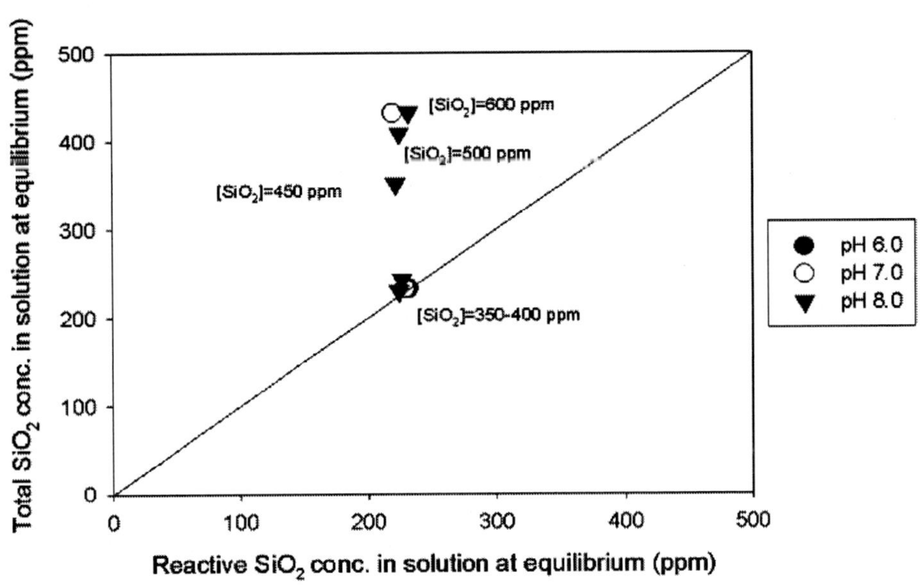

Fig 4.1.7 The solubility of amorphous SiO_2 at initial SiO_2 conc. 350-600 ppm and at 60°C

Data obtained for the silica thermodynamics suggested that the solubility of amorphous silica was largely independent of pH in the range of 6-8 (Fig 4.1.7, Appendix II). This is in agreement with results reported by Alexander et al. (1953) and other workers (Chan, 1989), who found the solubility of amorphous silica was relatively constant between pH 5-8. The solubilities of SiO_2 were found to increase from ~225 ppm at 60°C to ~295 ppm at 80°C in this work (Appendix IJ), consistent with values obtained in the literature (Rothbaum and Rohde, 1979).

4.1.2 COM-SiO2 Co-Precipitation in Binary Systems

4.1.2.1 Kinetics

Figure 4.1.8 presents the kinetics data for COM precipitation in the presence and absence of SiO_2. The results show the observed rate of COM precipitation was reduced by the presence of SiO_2 as indicated by the lower rate constants and higher n values (Fig 4.1.9). The extent of the reduction increased with increasing SiO_2 concentration (Fig. 4.1.9) (see Appendix IK and IL).

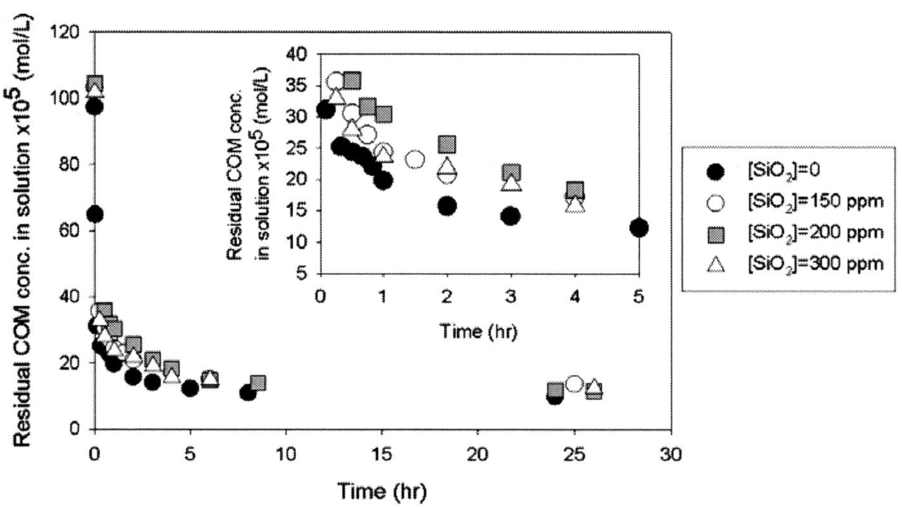

Fig. 4.1.8 COM precipitation kinetics in binary systems, at initial COM conc. 150 ppm, pH 8.0 and 60°C.

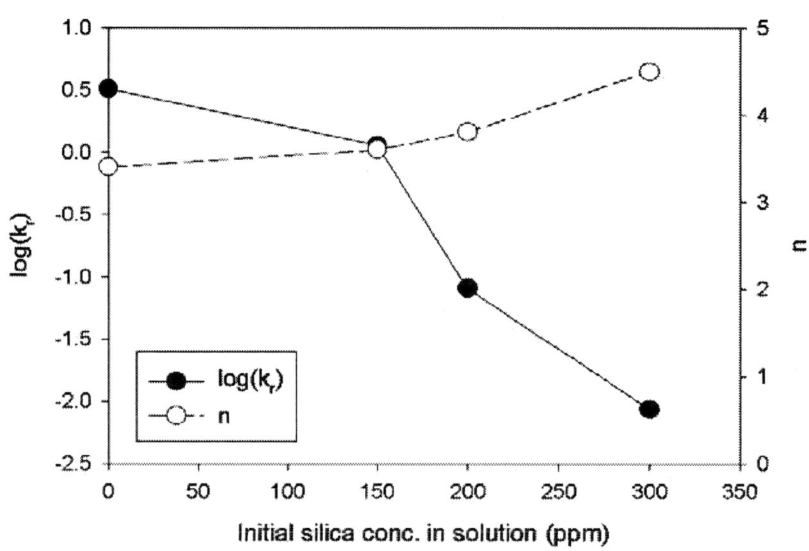

Fig 4.1.9 The changes in the rate constant (k_r) and order of reaction (n) for COM precipitation in binary systems as a function of initial SiO_2 conc., at initial COM conc. 150 ppm, pH 8.0 and at 60°C

The kinetics of SiO_2 polymerization in the presence of 20 and 200 ppm of COM are illustrated in Fig. 4.1.10 (see Appendix IM). The reactive SiO_2 concentrations decreased rapidly in the first 4 days and more than 90% of the equilibrium value was reached within 6-8 days for both binary systems. This indicates that the rate of silica polymerization from a low initial supersaturation level was enhanced by the presence of COM in comparison to that of pure SiO_2 solution where equilibrium was established in 14 days at 60°C. As reflected by the differences between total and reactive silica concentrations (Fig 4.1.10), colloidal particles of SiO_2 were found to occur in all binary mixtures in contrast to direct deposition of dissolved silica in SiO_2 single system.

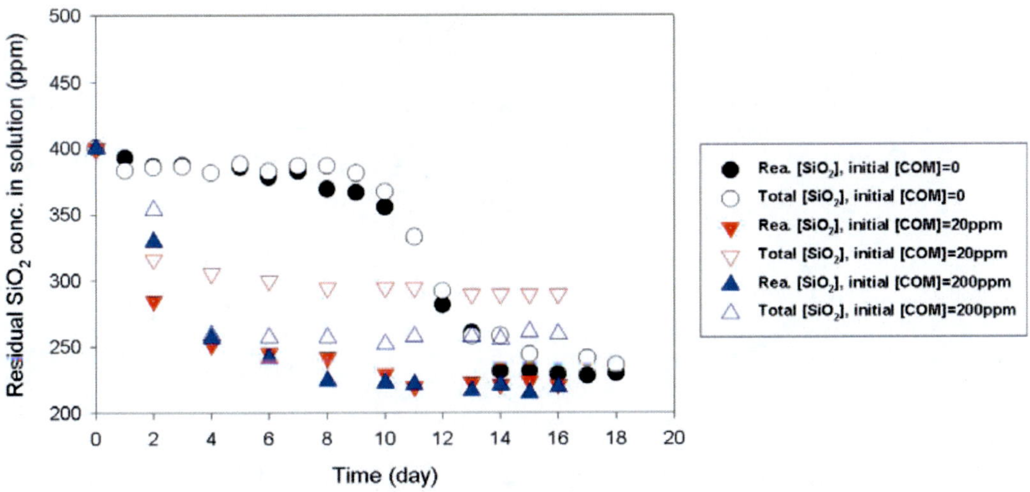

Fig. 4.1.10 The effect of COM on the kinetics of SiO_2 polymerization at initial SiO_2 conc. 400 ppm, initial pH 6.0 and 60°C

4.1.2.2 Thermodynamics

4.1.2.2.1 Three-Dimensional (3D) Solubility Surfaces

The precipitation thermodynamics of a single compound is usually studied in a two-dimensional (2D) plane, where the initial concentration of the compound is plotted against its final equilibrium (solubility) concentration in solution. However, when more than one precipitating species is present in solution, the evaluation, interpretation and representation of precipitation equilibria become more complicated since there is no control over interactive effects and the values of final equilibrium concentration of each component. In the case of binary system, the evaluation of precipitation thermodynamics leads to a series of 3D equilibrium (solubility) surfaces.

The 3D solubility surfaces in this study were obtained by plotting the initial concentrations of COM and SiO_2 on the x and y co-ordinates, respectively, against the solubility concentration of either one of them on the z co-ordinate. The computer software Matlab (version 6.0) was used for this purpose to connect random data points using interpolating mesh. It should be noted that this approach, however convenient, may produce surface irregularities, which may or may not truly reflect the behavior of a binary precipitation system.

4.1.2.2.2 COM Precipitation

Fig. 4.1.11 shows that the solubilities of COM generally increased with increasing SiO_2 concentration in the binary systems (Appendix IN) when compared with those obtained for single COM systems.

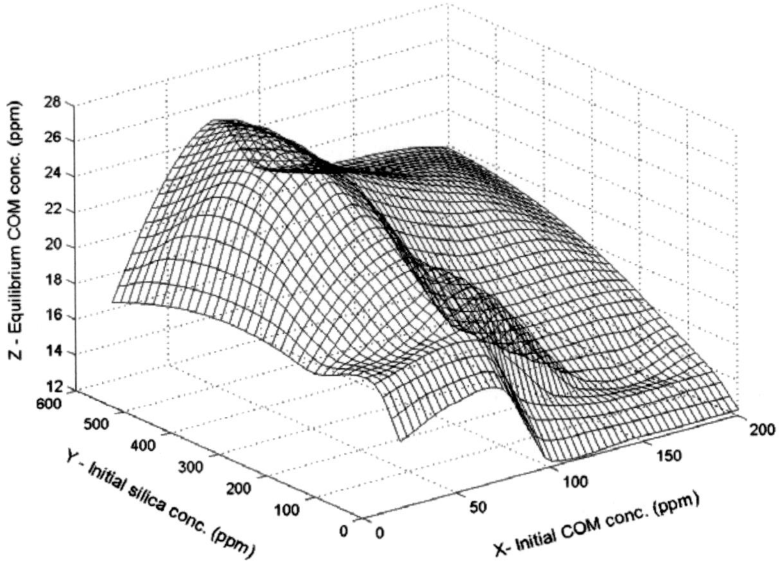

Fig. 4.1.11 Interpolated solubility surface for COM in binary system at initial pH 7.0 and at 60°C

To obtain two-dimensional solubility curves for COM at constant initial concentrations of SiO_2, the 3-D solubility surface can be "cut" by the constant SiO_2 concentration planes. Fig. 4.1.12 shows the outcome of the selected "cutting" through the surface in Fig. 4.1.11 by x-z planes representing various initial SiO_2 concentrations (experimental data shown by solid dots and values obtained from interpolation represented by lines). This approach better reveals the quantitative trends observed in the binary systems.

4.1.2.2.3 SiO_2 Polymerization

Figures 4.1.13 and 4.1.14 show that the solubilities of SiO_2 in the binary systems were virtually unchanged compared to the situation where SiO_2 was present alone (Appendix IN). However, the "Isocut" curves in Fig. 4.1.14 shows colloidal silica was formed at lower initial supersaturation levels in the binary systems than of the corresponding single system, i.e., colloidal silica formed at initial SiO_2 concentration of ~250 ppm in binary systems in comparison to greater than ~400 ppm in single system. These results are consistent with those obtained in the kinetics runs (Fig. 4.1.10).

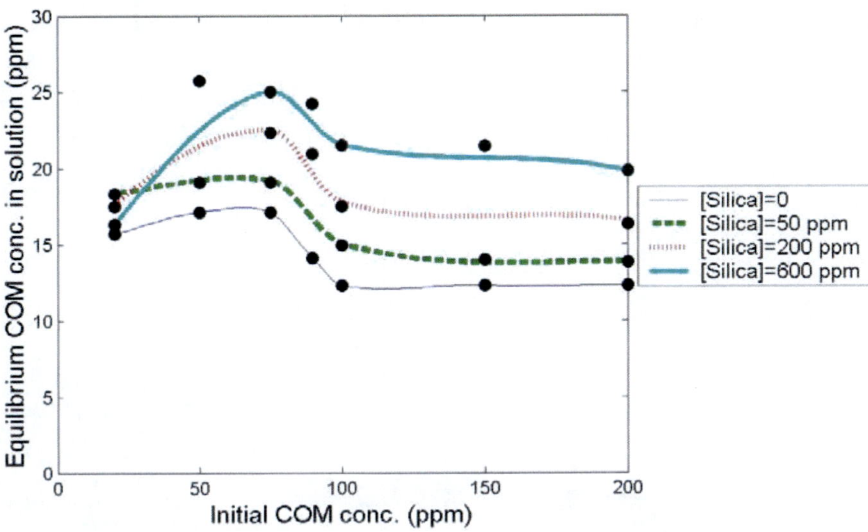

Fig. 4.1.12 Slicing through the surface in Fig. 4.1.11 by x-z planes representing initial SiO_2 conc. of 50, 200 and 600 ppm. The symbol • denotes the measured solubility data for COM.

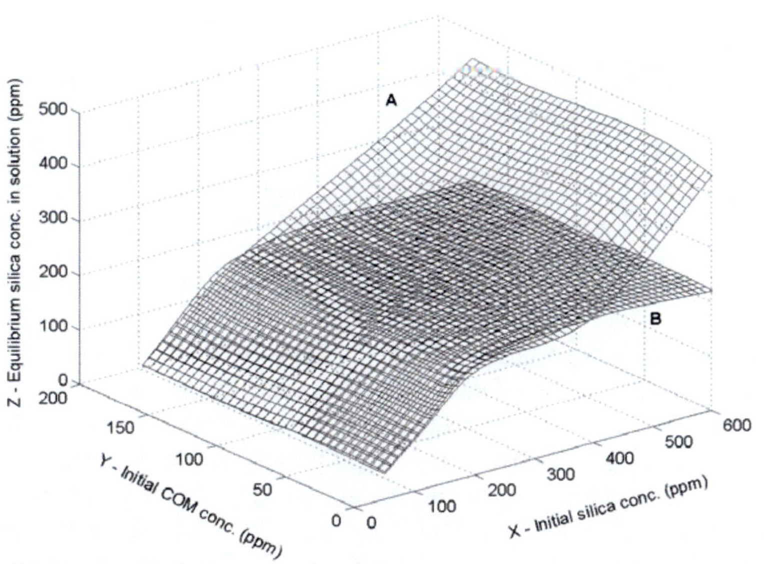

Fig. 4.1.13 Interpolated solubility surfaces for SiO_2 in binary system at initial pH 7.0 and 60°C, (A) total SiO_2 conc. (B) reactive SiO_2 conc.

4.1.2.3 Mechanisms

The effect of mutual interference between COM and SiO_2 was demonstrated on both kinetics and thermodynamics of COM precipitation and SiO_2 polymerization in the binary systems. A number of mechanisms may be proposed to explain the results described above. Iler (1979) demonstrated that SiO_2 was able to form a variety of soluble or insoluble silicates in the presence of non-alkali metals such as calcium and magnesium under alkaline conditions (pH $\geq$ 8.5). In another study, Iler (1975) examined the coagulation of colloidal silica by calcium ions. The author suggested that calcium ions act as bridging units between negatively charged SiO_2 particles. Carroll and Freeman (1954) investigated the binding affinities of alkaline earth and alkali elements for colloidal silicate and found a sequence of: Ca > Ba > Sr > K > Na > Li.

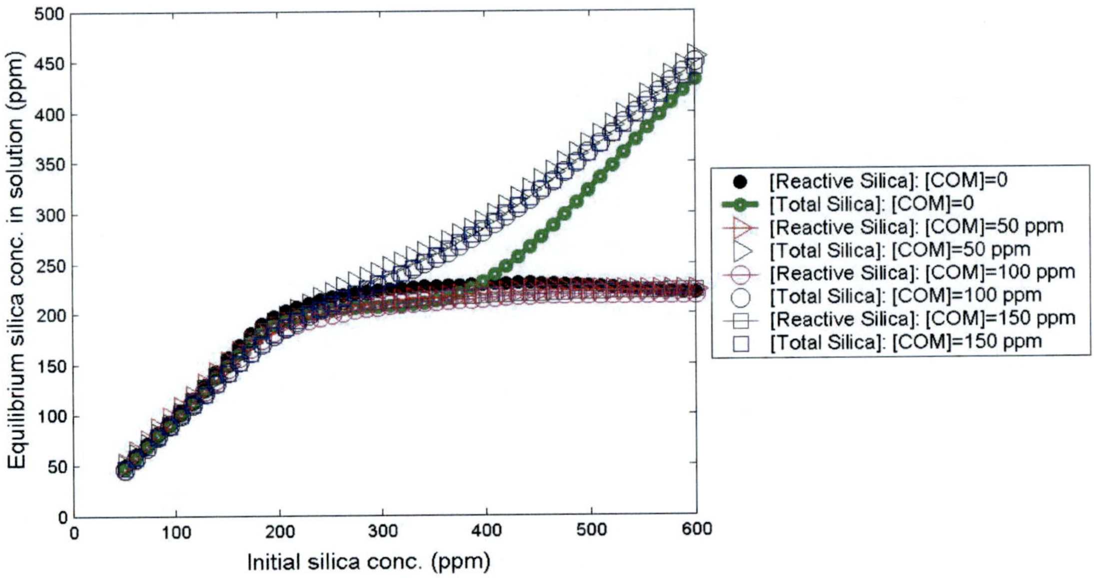

Fig. 4.1.14 Effect of COM on the solubilities of SiO_2 at initial pH 7.0 and 60°C, slicing through the surfaces in Fig. 4.1.13 by x-z planes representing initial COM conc. of 50, 100 and 150 ppm.

It is likely that similar mechanisms occurred during the interactions between COM and SiO_2 observed in this work. The reduced precipitation rate and increased solubility for COM might be considered a result of the formation of COM-SiO_2 species through complexation and bridging, etc., thereby maintaining high COM concentration in solution. Previous work on the crystallization of COM (Ryall, 1994) also showed that many agents could inhibit COM crystal growth by binding selectively to the COM crystal faces. Hence, the decreased precipitation rates for COM might possibly be attributed to the blocking of active growing sites of COM by the adsorption of SiO_2 ions or COM-SiO_2 species. This proposal is supported by the fact that the precipitation of COM itself is a surface controlled process. Furthermore, the potential phase transformation from COM into COT in the presence of SiO_2 may also be an important factor, as discussed in section 4.3.1.

Although the solubility concentration of SiO_2 was not altered by the presence of COM, there were ample evidences from the changes in the polymerization rate that the presence of COM affected the SiO_2 polymerization. Furthermore, the COM-SiO_2 complexes and COM may have acted as additional nucleation sites for SiO_2 polymerization, leading to high SiO_2 polymerization rates and the formation of colloidal SiO_2 at a lower level of initial SiO_2 supersaturation in the binary systems.

Based on the observation that the total SiO_2 concentrations dropped in the binary systems, it was also postulated that some soluble or colloidal SiO_2 may have co-precipitated with COM by specific adsorption and/or entrapment in the crystalline matrix of COM (Gill,

1998). An alternative explanation for the drop in total SiO_2 concentration may be associated with the coagulation and deposition of colloidal COM-SiO_2 complexes onto amorphous silica surfaces. Further experiments on crystal characterization are required to verify these assumptions.

4.1.3 Conclusions

Batch experiments have been conducted to examine the co-precipitation of amorphous silica and COM. The results show that the presence of SiO_2 in COM precipitation hindered the precipitation and increased the observed solubilities of COM, the extent of which increased with increasing initial SiO_2 concentration. The presence of COM in SiO_2 polymerization, however, increased the rate of SiO_2 polymerization, but had little effect on the solubility of SiO_2. Colloidal SiO_2 was formed at lower supersaturation levels in the binary systems than in the corresponding silica single system. The formation of COM-SiO_2 complexes was proposed to explain the co-precipitation of COM and SiO_2 in the binary systems.

4.2 COM-SiO_2 Co-Precipitation in Sugar Solutions

COM-SiO_2 co-precipitation in water has been shown to affect both the kinetics and thermodynamics of COM precipitation and SiO_2 polymerization due to the formation of a COM/SiO_2 complex and specific adsorption of silica particles onto COM crystal faces. The results indicated that while SiO_2 was likely to control the solubility of COM during co-precipitation, the kinetics of coprecipitation might be controlled by COM. In this section we investigate how the presence of sugar affects the co-precipitation thermodynamics and kinetics of COM and SiO_2. A batch experimental study using synthetic juice solutions was carried out for both single and binary foulant systems and the results compared with findings for water.

4.2.1 COM and SiO_2 Precipitation in Single Systems

Experiments involving the precipitation of COM and the polymerization of SiO_2 from single foulant sugar solutions were conducted for a range of sugar concentrations. COM showed higher precipitation rates in sugar solutions than that in water, with higher k and lower n values (Fig 4.2.1) according to the rate equation (eqn 4.1.1) (see Appendix IIA-B). The extent of increase in precipitation rate was clearly dependent on the sugar concentration (Fig 4.2.1). The reduced n values at high sugar concentrations seemed reasonable, as hindered diffusion were expected to be more significant with increasing viscosity in solution. The effect of sugar on the solubility concentrations of COM at different initial COM concentrations is shown in Fig. 4.2.2. The solubility of COM was decreased by the presence of sugar in solution in comparison to its solubility in water (i.e., from 13.6 ppm in water to 6.3 ppm in 40% sugar solution) and the extent of decrease increased with increasing sugar concentration (Appendix IIC). These observations corresponded well with the literature results, which showed reduced oxalate concentration with increasing sugar concentration

(Walford and Walthew, 1996). A solubility concentration peak at initial COM concentration of *ca.* 50 ppm was also found on the solubility curves of COM for all sugar solutions, in accordance with those observed for pure water, possibly due to the formation of colloidal particles achieved at this initial concentration level (section 4.1.1.2). The presence of sugar reduced the heights of these solubility peaks (Table 4.2.1), which gave an indication of the amount of colloidal COM species or a different type of calcium oxalate hydrate formed (Ulmgren and Radestrom, 1997).

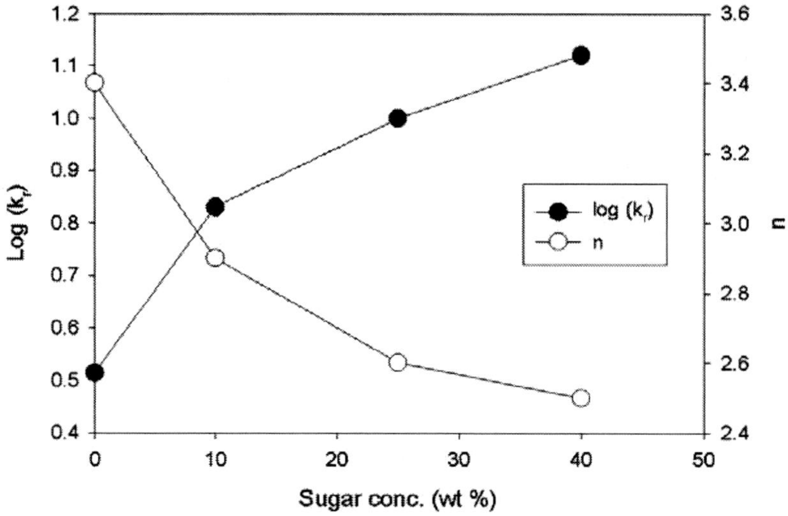

Fig 4.2.1 The changes in the rate constant (k_r) and order of reaction (n) for COM precipitation in single foulant sugar solutions as a function of sugar conc. at initial COM conc. 150 ppm, initial pH 8.0 and temp. 60°C; R^2 (correlation coefficient) > 0.995

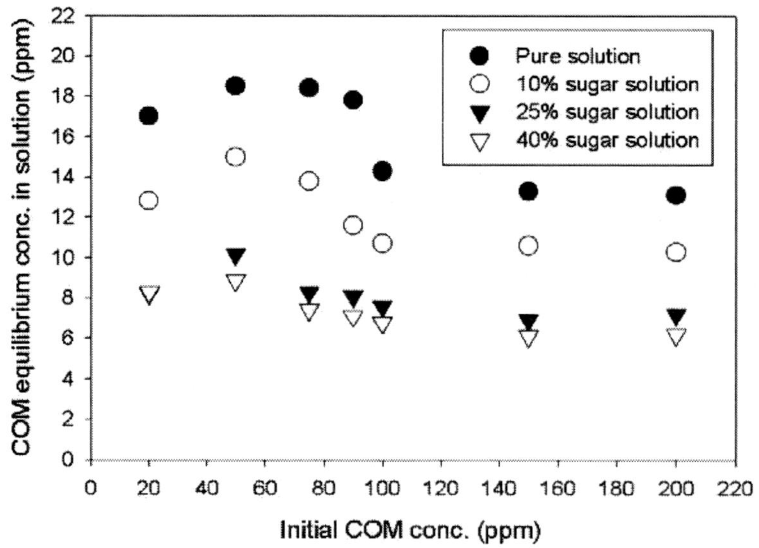

Fig 4.2.2 The effect of sugar on the solubility concentration of COM; initial pH 8.0 and temp. 60°C.

Table 4.2.1 COM solubility concentration data in water and sugar solutions

System designation	Peak conc. difference (ppm) *
Water	5.3
10% sugar solution	4.6
25% sugar solution	3.0
40% sugar solution	2.8

* Difference between the average equilibrium conc. value at high initial COM conc. and the peak value at initial COM conc. of 50 ppm

Similar kinetic enhancement was observed for silica polymerization in sugar solutions (Appendix IID). Fig 4.2.3 shows the solution concentration of reactive silica expressed in terms of supersaturation (SS) ratio as a function of reaction time. This allows comparison of polymerization rates based on different silica solubilities [The solubilities of silica in sugar solutions have been found to decrease with increasing sugar concentrations (Nandagopal and Ramamurthy, 1976; Poynton and Alexander, 1957), see Appendix II E for details]. Again, the rate of silica polymerization was increased by the presence of sugar, although in this case the extent of increase was much smaller at lower sugar concentrations (≤25%) (Table 4.2.2).

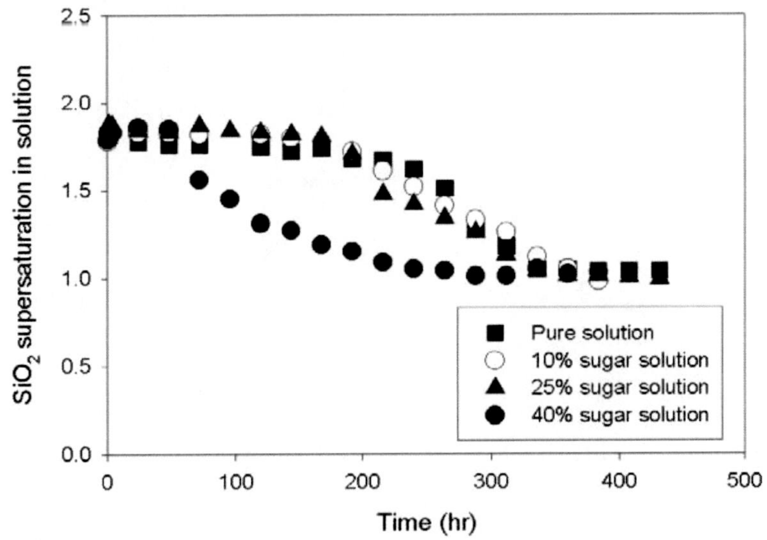

Fig 4.2.3 The kinetics of SiO_2 polymerization in single foulant sugar solutions; initial pH 6.0 and temp. 60°C.

Additional experimental evidence of the effect of sugar on SiO_2 polymerization thermodynamics was found by measuring the equilibrium reactive and total silica concentrations over a range of initial silica SS ratios (see Fig 4.2.4, Appendix IIF-G). In general, the differences between the equilibrium reactive and total silica concentrations imply the formation of colloidal (non-reactive) form of SiO_2 in solution (Sheikholeslami and Tan, 1999). Colloidal silica occurred in sugar solutions at initial silica SS ratio 1.8 or above, whereas little or no silica colloids could be found at initial silica SS of 1.8 in water. Thus the presence of sugar reduced the threshold initial silica SS for producing colloidal SiO_2 species from 2.0 to 1.8. Furthermore, the amounts of colloidal silica were reduced by the presence of sugar at initial silica SS of 2.0 (Fig 4.2.4).

Table 4.2.2 SiO_2 polymerization kinetics data in single foulant sugar solutions; initial SiO_2 SS ratio 1.8, pH 6.0 and temp. 60°C

System designation	Log (k_r)	n	Induction time (h)	$R^{2\ a}$
Water	-1.16	2.3	216	0.912
10% sugar solution	-1.10	2.1	192	0.920
25% sugar solution	-0.87	2.1	192	0.872
40% sugar solution	-0.32	1.8	48	0.980

[a] correlation coefficient

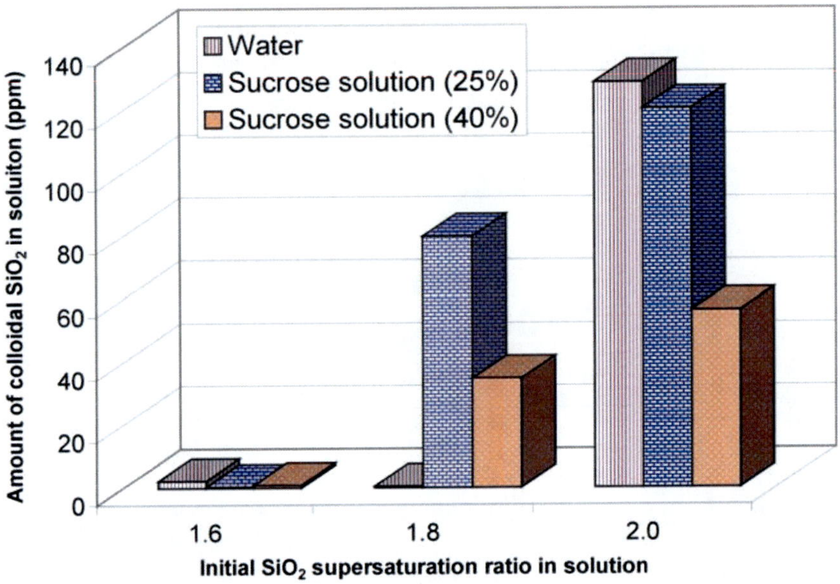

Fig 4.2.4 The amount of colloidal SiO_2 in water and sugar solutions at equilibrium, initial pH 8.0 and temp. 60°C

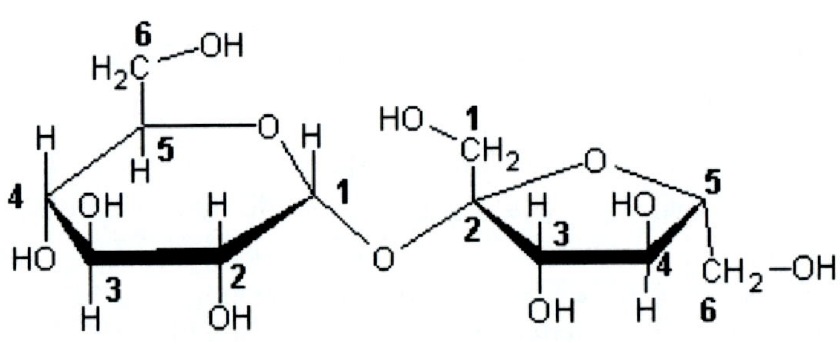

Fig 4.2.5 Sucrose molecule.

The above results demonstrated that the presence of sugar (more specifically sucrose) in solution had considerable effect on the kinetics and thermodynamics of both COM precipitation and SiO_2 polymerization. This may be due to several factors including complex formation with calcium oxalate and silica species by sucrose molecules (Fig 4.2.5), via coordination to the hydroxyl groups of sucrose without deprotonation of hydroxyl groups (Kollonitsch, 1970; Pannetier et al., 2001; Sahai and Tossell, 2000), as supported by the little changes in solution pH during reaction, and the availability of water molecules. The sucrose complexes promoted COM precipitation and SiO_2 polymerization by providing additional nucleation and growth sites, resulting in the formation of colloidal silica at a lower initial silica SS ratio. These polymeric compounds might continue to aggregate the colloidal species of SiO_2 by further complexation or adsorption. Therefore the amounts of SiO_2 colloids would be generally lower in sugar solutions than those found in water (see Fig 4.2.4). Additionally, the intermolecular associations between sucrose and water (Starzak and Peacock, 2000) may have influenced the solubilities of COM and SiO_2 by reducing the number of water molecules available for the solvation of solutes (COM and SiO_2) (Reichardt, 1988).

4.2.2 COM-SiO_2 Co-Precipitation in Binary Systems

4.2.2.1 Kinetics

The results obtained in the kinetic studies of COM precipitation from binary foulant solutions containing 25% sugar are shown in Fig 4.2.6. These are representative of the kinetic behavior of COM precipitation in sugar solutions of various concentrations (Appendix IIH). It was observed that COM precipitation was accelerated by the presence of SiO_2 in sugar solution, in contrast to the situation with binary foulant mixtures in water (section 4.1). The difference might be associated with the formation of various polymeric species involving sugar and SiO_2 as previously described. At high SiO_2 concentrations, these colloids might flocculate and form highly gel-like particles, which not only act as new growth centers, but also induce aggregation by trapping COM crystallites in the matrix, as has been demonstrated for other biological macromolecules (Campbell et al., 1989; Hess et al., 1991). The order of reaction was found to decrease (from 3.0 to 2.2) with increasing SiO_2 concentration, a trend similar to that observed for the effect of sugar concentration on the precipitation kinetics of COM (see Fig 4.2.1 and Table 4.2.2). This again could be attributed to a hindered diffusion effect as a result of increased polymeric complex formation between SiO_2 and sucrose, which might induce an increase in solution viscosity.

Fig 4.2.7 shows that COM precipitation rate also increased with increasing sugar concentration in COM-SiO_2 binary foulant sugar solutions (Appendix III-J), consistent with the effect of SiO_2 (Fig 4.2.6) and proposed mechanisms.

Figs 4.2.8 and 4.2.9 show the effect of COM on the rate of SiO_2 polymerization at sugar concentrations of 25 and 40%, respectively (Appendix IIK-L). In a less concentrated sugar solution (Fig 4.2.8), the rate of SiO_2 polymerization was seen to be significantly increased by the presence of COM when compared to that of the corresponding single foulant sugar solution, although higher COM concentrations over 50 ppm caused little further increase in polymerization rate constant and the order of reaction, which was largely between 1-2 (Table 4.2.3). A Similar effect of COM was also observed for the binary foulant mixtures in water (see Fig 4.1.10).

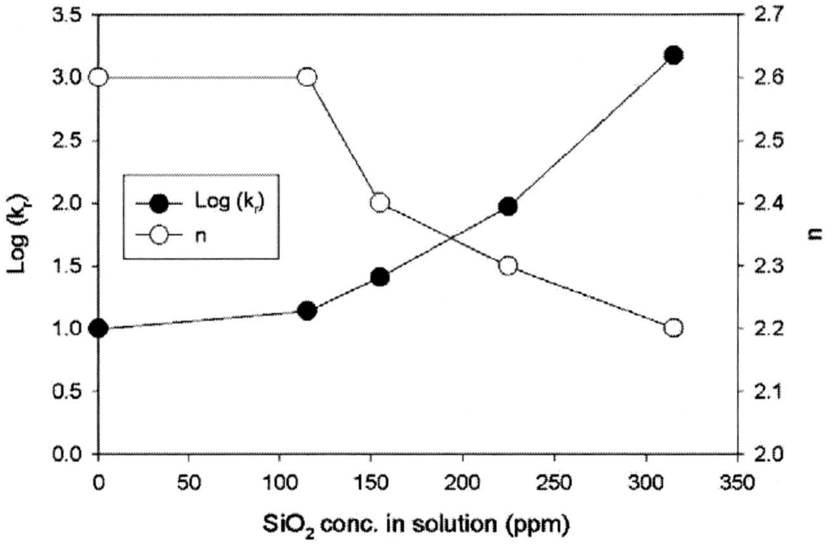

Fig 4.2.6 The changes in the rate constant (k_r) and order of reaction (n) for COM precipitation in binary foulant sugar solutions as a function of initial SiO_2 conc. measured at 25% sugar conc. (wt %), initial COM conc. 150 ppm, pH 8.0 and temp. 60°C; R^2 (correlation coefficient)>0.995

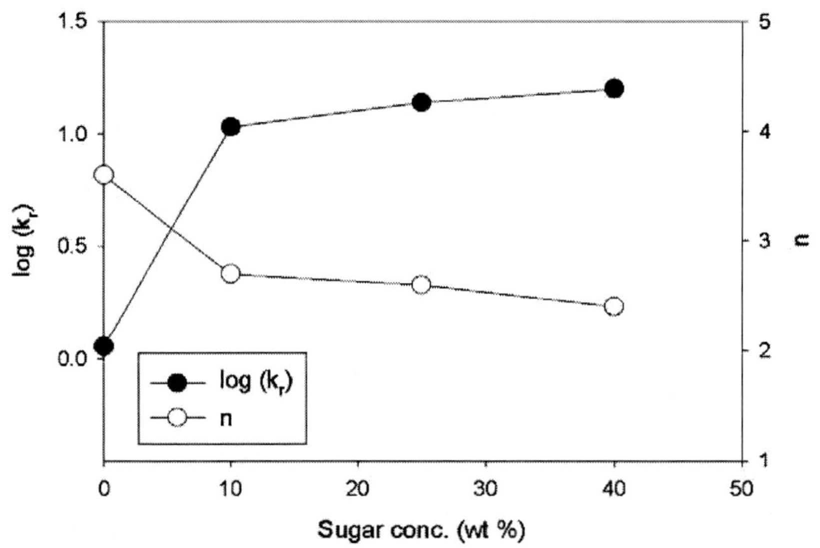

Fig 4.2.7 The effect of sugar conc. on COM precipitation kinetics in binary foulant sugar solutons at initial [COM]=150 ppm, initial SiO_2 SS 0.7, pH 8.0 and 60°C

On the other hand, the rate of SiO_2 polymerization in a more concentrated sugar solution (Fig 4.2.9) was slightly increased at low COM concentration (50 ppm) but then decreased with increasing COM concentrations. As shown in Table 4.2.3, the induction period increased by a factor of 32 when the COM concentration was raised from 50 ppm to 150 ppm. Similar trend was observed for the effect of sugar on SiO_2 polymerization (Fig 4.2.10): the rate of SiO_2 polymerization showed a slight increase at low sugar concentrations (≤ 33%), followed by a significant reduction with further increase in the sugar concentration (Appendix IIM). Therefore it appears that SiO_2 polymerization in binary foulant sugar solutions was promoted at either low COM or sugar concentration (or both), but could be inhibited at high concentrations of both COM and sugar.

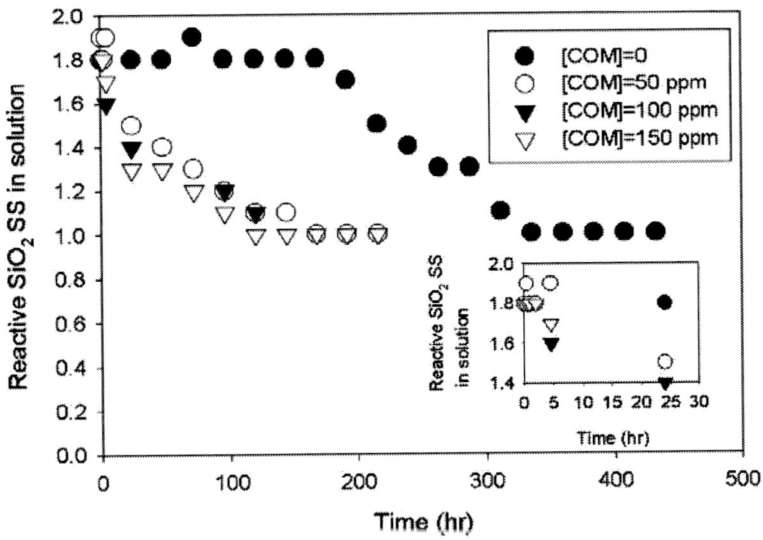

Fig 4.2.8 The effect of COM on the kinetics of SiO_2 polymerization in binary foulant sugar solutions containing 25% sugar at initial pH 6.0 and 60°C.

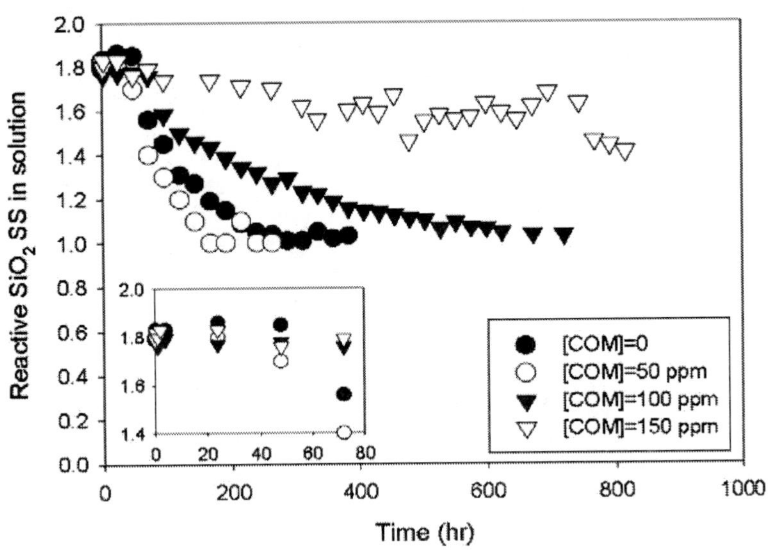

Fig 4.2.9 The effect of COM on the kinetics of SiO_2 polymerization in binary foulant sugar solutions containing 40% sugar at initial pH 6.0 and 60°C.

Table 4.2.3 Summary of SiO_2 polymerization kinetics data * in single and binary foulant sugar solutions over a range o sugar concentrations

System designation	25% sugar solution				40% sugar solution			
	log (k_r)	n	Induction time (h)	R^2	log (k_r)	n	Induction time (h)	R^2
[COM]=0 ppm	-0.87	2.1	168	0.872	-0.32	1.8	48	0.980
[COM]=50 ppm	0.36	1.5	5	0.996	-0.07	1.5	24	0.968
[COM]=100 ppm	0.38	1.2	2	0.997	-0.49	1.9	72	0.991
[COM]=150 ppm	0.40	1.5	2	0.998	n/a **	n/a	770	-

* measured at initial SiO_2 SS ratio 1.8, pH 6.0 and temp. 60°C
** Not available

The role of COM or sugar as either a polymerization promotor or inhibitor of silica in sugar solutions may be attributed to the following:

- The formation of sucrose adducts/complexes with SiO_2 and COM, and
- Subsequent adsorption of COM-sucrose species onto SiO_2 polymeric complexes (and vice versa) (Baudin et al., 1990; Esumi and Matsui, 1993).

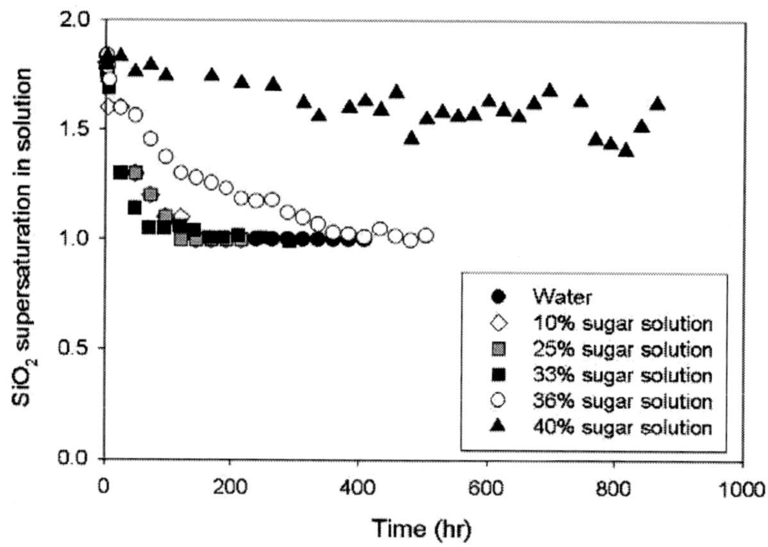

Fig. 4.2.10 Effect of sugar concentration on the kinetics of silica polymerization in binary-foulant sugar solutions at initial COM concentration 150 ppm, initial pH 6.0 and 60°C

The presence of low concentration of either COM or sugar accelerated SiO_2 polymerization due to a surface charge neutralization of colloidal SiO_2 (Callejas-Fernandez et al., 1991; Schwarz et al., 2000) by the incorporation of COM – sucrose complexes. This hypothesis is supported by the result that the zeta-potential of a COM-SiO_2 composite deposit was lower than that of pure SiO_2 particles (see section 4.5). At high sugar and COM concentrations, which would result in an increasing amount of COM-sucrose complexes formed in solution, silica polymerization was suppressed due to an over-balancing of surface charges of silica particles by the adsorbed COM-sucrose complexes. This led to the stabilization of silica colloids. Moreover, the COM – sucrose macromolecules adsorbed might also stabilize the colloidal silica by steric forces (Bauer et al., 1999), especially at high COM concentrations (e.g., 150 ppm) when sucrose complexes mainly consisted of branched or cross-linking structures (Pannetier et al., 2001).

4.2.2.2 Thermodynamics

To evaluate the precipitation equilibrium of a binary foulant system, three-dimensional (3D) precipitation equilibrium surfaces were constructed by plotting the solubility concentration of either one of the precipitating species as a function of the initial concentrations of two species (see section 4.1.2.2.1). Figs 4.2.11 and 4.2.12 show examples of COM and SiO_2 solubility surfaces, respectively, for a given sugar concentration of 25% (Appendix IIN); similar results were obtained from different sugar solutions. There are lumps and irregularities on the surfaces possibly because of experimental errors associated with those data represented in space. To quantitatively assess the interfering effect between two species, a solubility surface were cut by selected "iso-concentration" planes at chosen initial concentrations of the interfering compound, yielding a series of two-dimensional solubility curves for the main compound as a result of the increasing presence of the interfering compound. The solubility curves in Figs 4.2.13 and 4.2.14 represent the respective "iso-cuts" of COM and SiO_2 solubility surfaces from Figs 4.2.11 and 4.2.12.

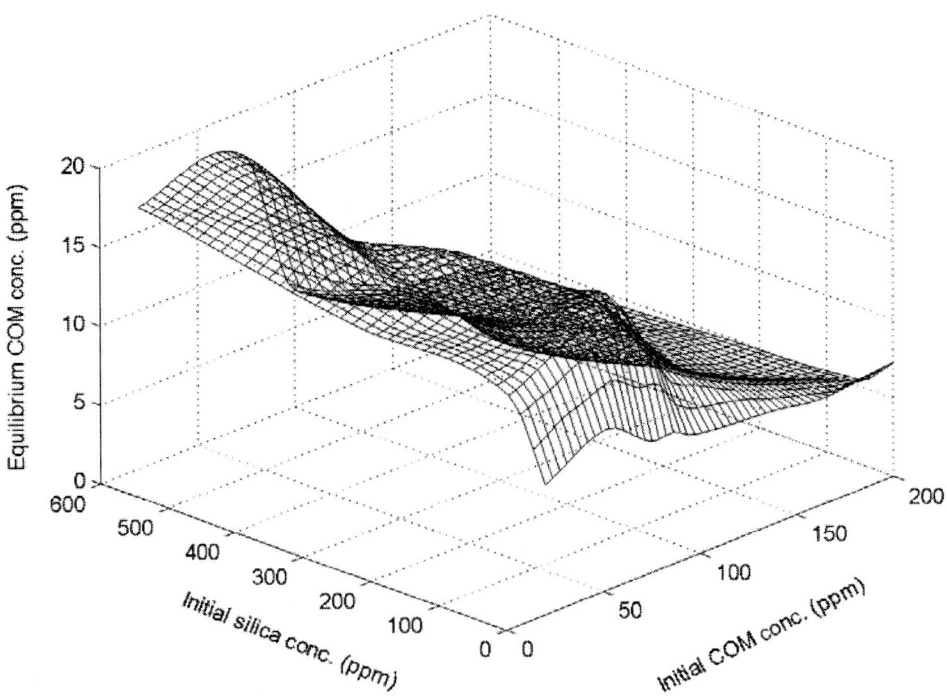

Fig 4.2.11 Interpolated solubility surface for COM in binary foulant sugar solutions at sugar conc. 25%, initial pH 8.0 and at 60°C.

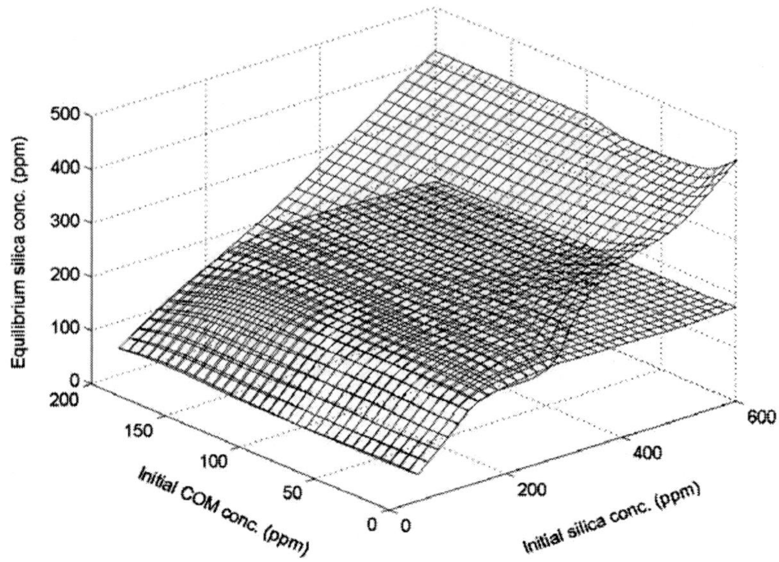

Fig 4.2.12 Interpolated solubility surfaces for SiO_2 in binary foulant sugar solutions at sugar conc. 25%, initial pH 8.0 and 60°C. (A) total SiO_2 conc. (B) reactive SiO_2 conc.

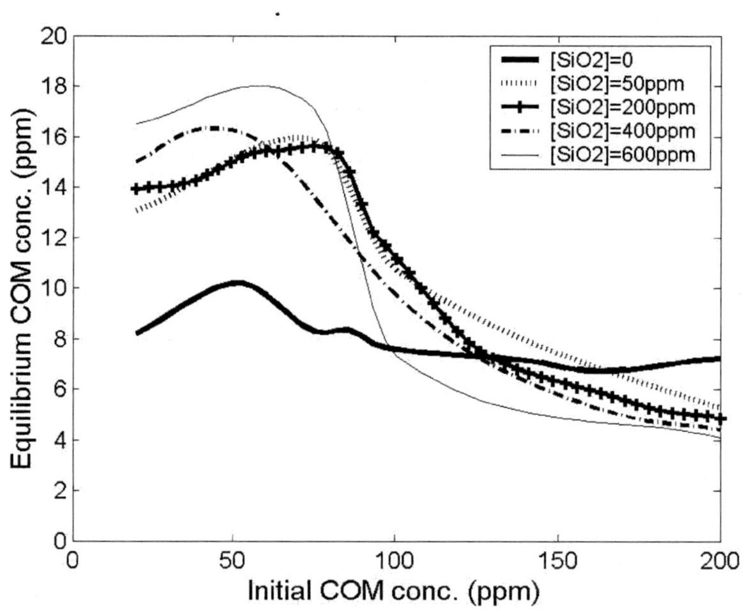

Fig 4.2.13 Effect of SiO_2 on the solubilities of COM at 25% sugar conc., initial pH 8.0 and 60°C, slicing through the surface in Fig. 4.2.11 by x-z planes representing initial SiO_2 conc. of 50, 200, 400 and 600 ppm, respectively.

Fig 4.2.13 shows that the equilibrium concentrations of COM in the binary foulant sugar solutions were much higher at low initial COM concentration range (<100 ppm) but sharply decreased to be slightly lower at high initial concentrations of COM in comparison to those in the single foulant sugar solution. As a result, the peak heights at ~50 ppm COM were markedly higher than that of COM single system. The solubility of COM also decreased with increasing sugar concentration (Table 4.2.4). This was in agreement with results obtained from pure COM sugar solutions (Fig 4.2.2), possibly as a result of solvation of sucrose molecules (see section 4.2.1). The presence of COM in sugar solution, on the other hand, was found to lower the threshold initial silica SS levels (from ~1.7-1.8 to ~1.3; see Fig 4.2.14) for the formation of colloidal silica species but had little effect on the solubility of SiO_2, showing good correspondence to the effects of COM observed in water (see section 4.1).

Table 4.2.4 The effect of sugar conc. on COM solubility in binary foulant sugar solutions at initial SiO_2 SS 1.8, pH 8.0 and 60°C

Sugar conc. (%)	COM solubility conc. (ppm) *
0	18.0
10	10.0
25	4.7
40	4.0

* average equilibrium conc. at high initial COM conc.

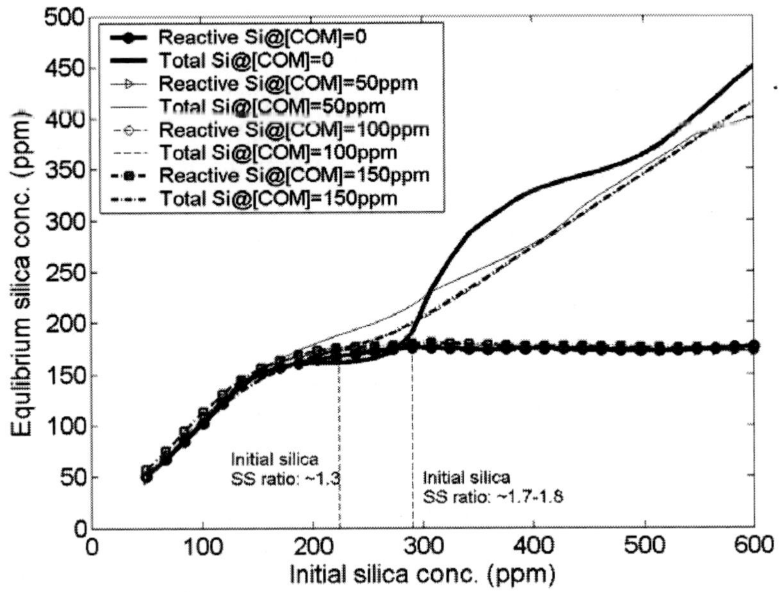

Fig 4.2.14 Effect of COM on the solubilities of SiO_2 at 25% (wt %) sugar conc., initial pH 8.0 and 60°C, slicing through the surfaces in Fig. 4.2.12 by x-z planes representing initial COM conc. of 50, 100 and 150 ppm.

The variations in the COM equilibrium concentrations observed at different COM SS levels in the binary foulant sugar solutions might be accounted for by the presence of COM-SiO_2-sucrose complexes of colloidal form at low initial COM supersaturations, which were retained in the sample solutions after the membrane filtration (0.22 micron). With increasing initial COM concentrations, these colloidal species shifted toward larger particles because of additional particle aggregation and deposition, leading to a reduction in the observed solubility of COM. The drop in the threshold silica SS might again be due to the complex formation between COM, SiO_2 and sucrose, which provided additional active centers for heterogeneous nucleation of silica polymerization.

The above experiments demonstrate that sugar largely controls the thermodynamics and kinetics of COM and SiO_2 coprecipitation in sugar solutions, although the effect of COM on coprecipitation kinetics is also important. Thus composite fouling data obtained in water do not necessarily reflect the practical scaling behavior/mechanisms in sugar mill evaporators, which are significantly affected by the presence of sugar in solution. These results have practical implications for future development of new pre-treatment, cleaning and control strategies and of a predictive model to be applied for composite fouling of COM and SiO_2. Further work is required, using instrumental analysis to elucidate the proposed mechanisms of co-precipitation between COM and SiO_2 in sugar media. The effects of sugar in solution on the rate of composite fouling and the characteristics of composite scales will also be examined, in a dynamic system which simulates the thermo-hydraulic conditions of sugar mill evaporators.

4.2.3 Conclusions

This work has shown that the presence of sugar affected the kinetics and thermodynamics of COM precipitation and SiO_2 polymerization in single foulant systems. The rates of COM precipitation and SiO_2 polymerization were found to depend on the concentration of sugar in solution, with the highest rates obtained for the most concentrated sugar solution. The solubilities of COM and SiO_2 were found to decrease with increasing sugar concentration. The amounts of colloidal SiO_2 particles formed in the sugar solutions were lower than those in water. The threshold initial silica SS for colloidal silica to form was also slightly lower in sugar solutions than that required in the absence of sugar.

The results of COM and SiO_2 co-precipitation experiments in binary foulant sugar solutions have established, for the first time, that the presence of SiO_2 or sugar accelerated the precipitation of COM. However, COM or sugar served to either promote or inhibit SiO_2 polymerization, depending on the initial COM and sugar concentrations. SiO_2 polymerization rate was increased at low concentration of either COM or sugar, but was reduced at high concentrations of both COM and sugar. On the other hand, the presence of SiO_2 or sugar decreased the solubilities of COM, and led to an increase in the solubility peak height at an initial COM concentration of around 50 ppm. The presence of COM further decreased the threshold silica SS requirement for colloidal silica formation, although it had little effect on the solubility of silica in sugar solution. The overall interactions between sugar, COM and SiO_2 in the single and binary foulant systems have been attributed to adduct formation between COM, SiO_2 and sucrose, sucrose hydration and adsorption of COM-sucrose complexes onto the surface of polymeric SiO_2 species.

4.3 Characterization of CaOx Crystals and Composite Deposits Formed in Water and Sugar Solutions

Batch tests evaluating the thermodynamics and the kinetics of CaOx-SiO_2 co-precipitation in water and sugar solutions have been conducted to provide a basis for an understanding of the mechanisms responsible for the composite fouling of CaOx and SiO_2 in sugar mill evaporators. The results showed that the presence of SiO_2 in supersaturated solutions of CaOx led to a decrease in the rate of CaOx precipitation and an increase in the solubility of CaOx. The presence of CaOx in supersaturated solutions of SiO_2 enhanced the SiO_2 polymerization and lowered the supersaturation level required for colloidal SiO_2 to occur as a result of the formation of polymeric CaOx-SiO_2 complexes. Based on the order of reaction and activation energy values of CaOx and SiO_2 single systems, it was proposed that the co-precipitation of CaOx and SiO_2 was likely to be a surface-controlled process.

The kinetics and the thermodynamics of CaOx precipitation, SiO_2 polymerization and CaOx-SiO_2 co-precipitation in sugar solutions were found to be markedly different from those observed in water. The presence of sugar accelerated both the precipitation of CaOx and polymerization of SiO_2 and reduced their solubilities in single foulant (i.e., scale-forming component) solutions. In the binary foulant sugar solutions, the presence of SiO_2 enhanced the precipitation of CaOx, whereas CaOx either promoted or inhibited SiO_2 polymerization, depending on the initial CaOx and sugar concentrations. A number of mechanisms were proposed to explain the CaOx-SiO_2 co-precipitation behaviour observed in sugar solutions.

In this section, a combination of instrumental techniques including SEM-EDS, XRD and FT-IR spectroscopy have been employed to examine the crystallographic and morphological changes of CaOx and CaOx/SiO_2 composites formed in water and sugar solutions. The effects of the initial CaOx, SiO_2 and sugar concentrations on the phase composition and morphology of CaOx were studied over a range comparable to those used in the batch co-precipitation tests. Results of this work were expected to provide complementary information to previous studies on CaOx/SiO_2 co-precipitation, and will help establish the composite fouling mechanisms of CaOx and SiO_2 in sugar mill evaporators.

4.3.1 Analysis of Deposits Formed in Aqueous Solutions

Fig 4.3.1 shows the XRD spectra of CaOx and CaOx-SiO_2 composites formed in water. COM was found to be the only crystalline phase present in CaOx (Fig 4.3.1A) and the composite obtained in the presence of silica SS of 1.8 or lower (Fig 4.3.1B) (Kontoyannis et al., 1997). At higher SiO_2 SS of 2.3 and 3.0, the XRD data indicated a phase transition from COM to COT in the composite deposits (Fig 4.3.1C and D). This result was similar to findings reported previously for silica co-precipitation with calcium carbonate, which showed that the presence of SiO_2 inhibited the formation of aragonite in favour of the formation of calcite (Kitano et al., 1979).

The electron micrographs of CaOx and CaOx/SiO_2 composites were taken to observe the morphological features of crystals under different co-precipitation conditions. Generally, CaOx crystallises in various shapes and forms according to the degree of hydration it acquires: as monoclinic prismatic monohydrate, tetragonal dihydrate or triclinic and needle-shaped trihydrate (Brecevic et al., 1986; Grases et al., 1990; Heijnen, 1982; Singh et al., 1988) (Fig 4.3.2).

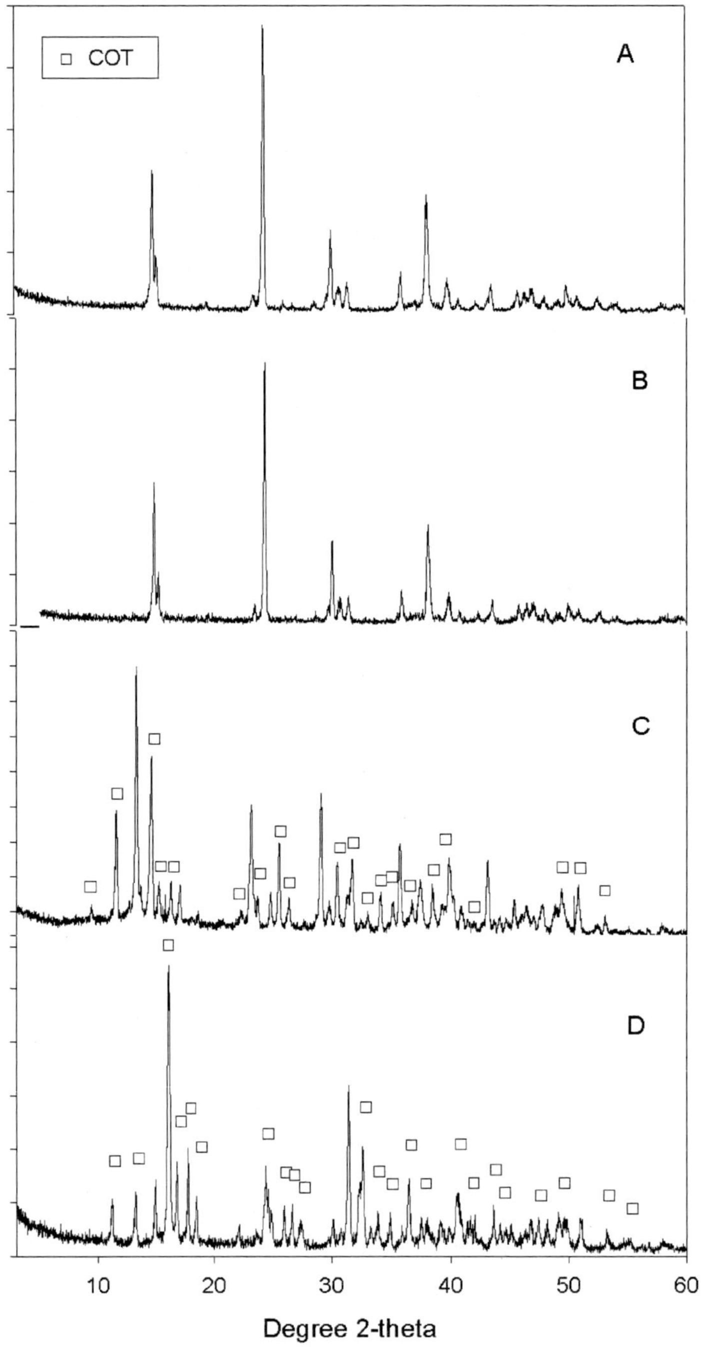

Fig. 4.3.1. XRD graphs of crystals formed in water at initial CaOx conc. of 175 ppm, initial pH of 8.0 and 60°C; (A) single CaOx system; binary CaOx/SiO_2 systems at initial silica SS of (B) 1.8, (C) 2.3 and (D) 3.0

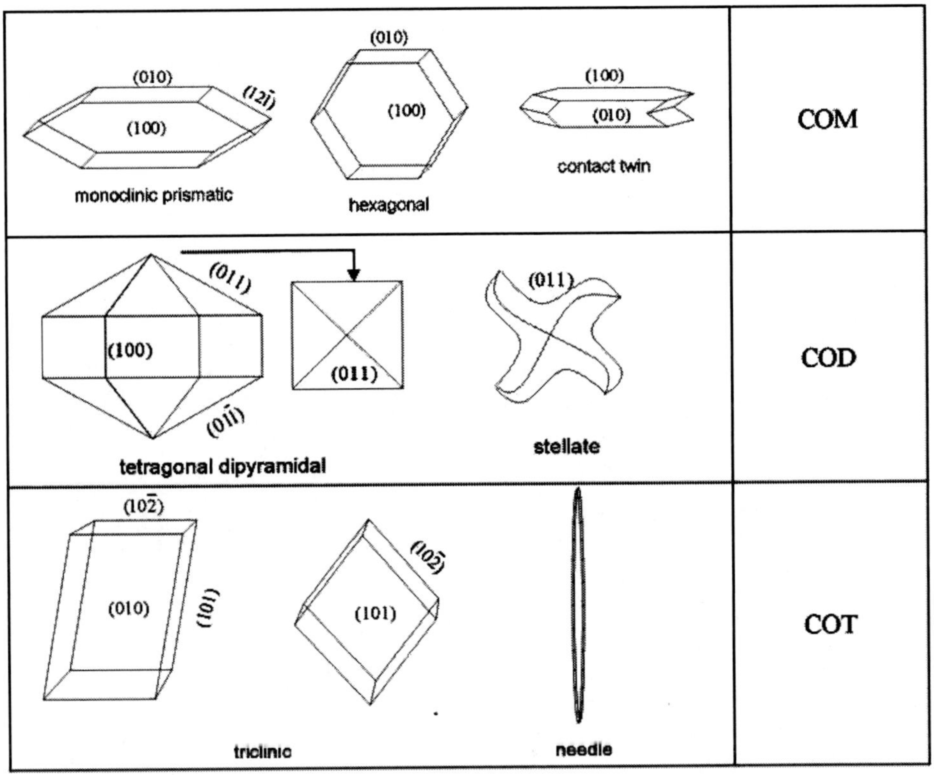

Fig. 4.3.2 Crystal habits of COM, COD and COT obtained from water and sugar solutions (crystal face notation system adopted from Tazzolli (Millan, 2001)

Figs 4.3.3A and 4.3.3B show that the precipitates obtained from CaOx single system and the binary mixture at a low initial SiO_2 SS contained aggregates of monoclinic COM crystals, with contact twinning [attachment of two individual crystals along (100) plane] being more prevalent in the binary mixture (Fig 4.3.3B). Fig 4.3.3C and 4.3.3D show that the presence of high SiO_2 SS (≥2.3) promoted the formation of COT (Fig 4.3.2) with COM crystals barely visible, consistent with the XRD spectra (Fig 4.3.1). The COM crystals formed in the CaOx-SiO_2 binary system were found to be much thinner along (100) faces (see Fig 4.3.2) than those seen in CaOx single system. Many COT crystals (Fig 4.3.3C) appeared to originate out of one centre, indicating a lack of available COT nuclei (Tunik et al., 1996) during growth. Thus, the nucleation of COT in water may also be inhibited by SiO_2 – albeit to a less extent than that of COM.

Fourier transform infrared (FT-IR) spectroscopy was used to analyse the molecular structure of sample components. The IR spectra of CaOx and composites exhibited peaks around 1628 cm^{-1}, 1317 cm^{-1} (symmetrical and antisymmetrical oxalate C=O bond stretching), 781 cm^{-1} (O-C-O in-plane bending) and 517 cm^{-1} (O-C-O in-plane rocking), characteristic of COM (Babic-Ivancic et al., 1985) (Fig 4.3.4). The OH stretching of coordinated water molecules of COM was shown by the broad absorption bands between 3000-3500 cm^{-1} in all samples. The water libration bands of the composites (660 and 584 cm^{-1}) were slightly shifted to be lower than those of the CaOx (667 and 594 cm^{-1}), probably due to an increase in the number of water molecules in the crystal structure, i.e., the presence of

COT (Babic-Ivancic et al., 1985). This observation was in contrast to the results of the XRD and SEM imaging in which COT was found to be the main component of the composite formed at high SiO_2 SS of 2.3 and 3.0. As shown later, this change was merely an artifact of the way the samples were prepared. Another weak band at 1113 cm^{-1} due to Si-O-Si stretching vibration of amorphous silica was also present in the spectrum of composite deposit (Fig 4.3.4B) (Bellamy, 1975).

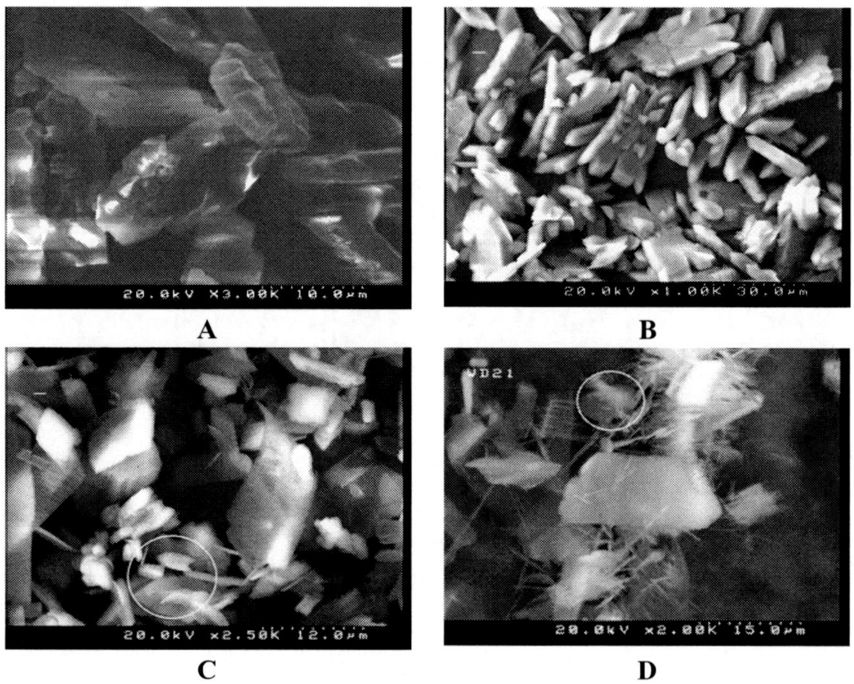

Fig. 4.3.3. SEM images of crystals from water at initial CaOx conc. 175 ppm; (A) single CaOx system and binary CaOx/SiO_2 systems at initial SiO_2 SS of (B) 1.8, (C) 2.3 and (D) 3.0, (white circle denotes COM crystal)

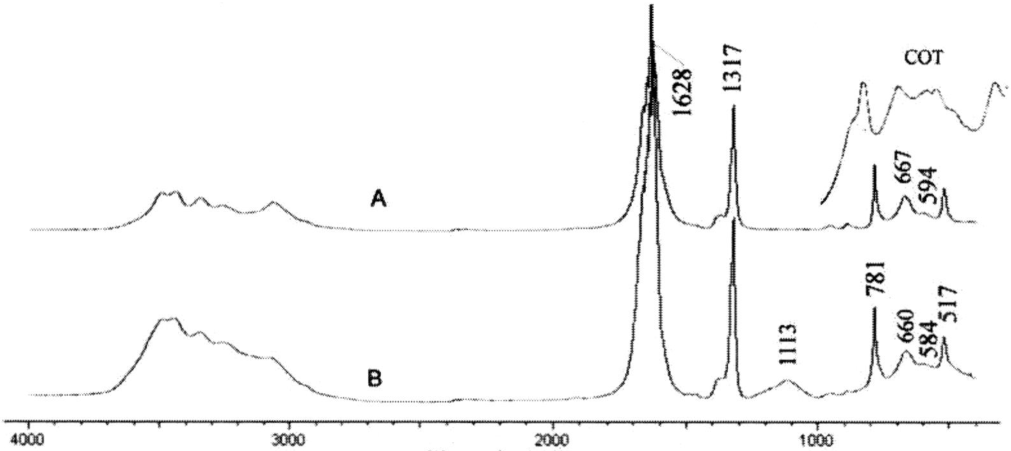

Fig. 4.3.4 FT-IR spectra of (A) CaOx and (B) all composites obtained from water

4.3.2 Analysis of Deposits Formed in Sugar Solutions

XRD patterns of CaOx precipitated from sugar solutions of increasing concentrations (Fig 4.3.5) showed a gradual shift from COM to COD. COD was also identified as the dominant phase in all CaOx/SiO_2 composite materials formed in sugar solutions along with traces of COM, COT and amorphous silica (indicated by a broad peak or 'halo' at 2θ value of ~12° (Nakamura et al., 1989), see Fig 4.3.6). Thus the presence of sugar in CaOx solution has induced the phase transformation, and a trend of increased transformation in single foulant sugar solutions was evident with increasing sugar concentrations.

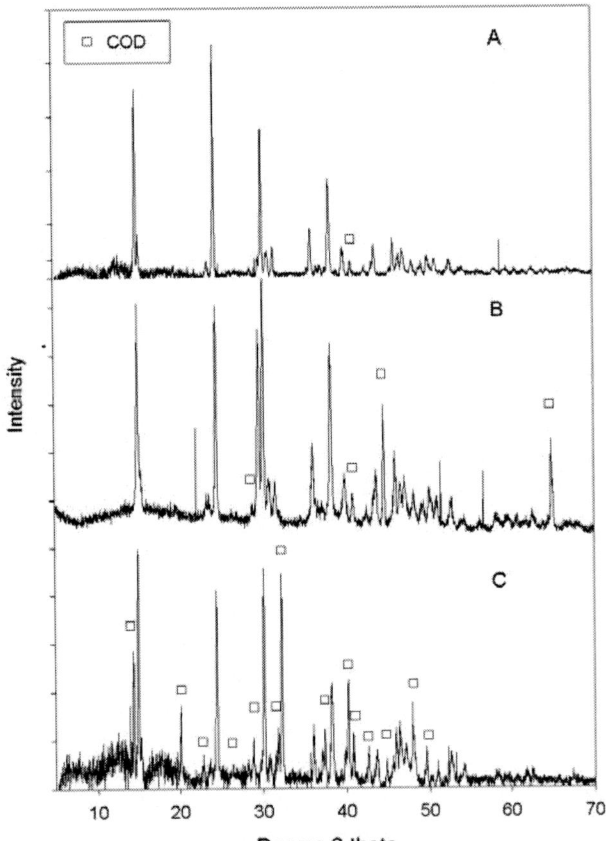

Fig. 4.3.5. XRD spectra of CaOx crystals formed at an initial CaOx conc. of 130 ppm in sugar solutions of (A) 10%, (B) 25% and (C) 40%

The changes in the size and morphology of CaOx crystals as a function of sugar concentration are presented in Figs 4.3.7A - C. Monoclinic COM crystals were shown to be the major component at low and intermediate sugar concentrations (i.e., 10% and 25%, Figs 4.3.7A/B), with a few COD bipyramids (Fig 4.3.2) appearing in 25% sugar solution. At a high sugar concentration of 40% (Fig 4.3.7C), COD crystals became the main component, but were considerably smaller and more numerous than those formed in 25% sugar solution, indicating an enhanced nucleation of COD. Hence the sucrose molecules may have a much less inhibitive effect on COD than that for COM or even act as a template for the heterogeneous nucleation of COD. As the phase composition moved further into COD dominance, the growth of COM crystal was gradually inhibited and it became thinner and smaller with increasing sugar concentration.

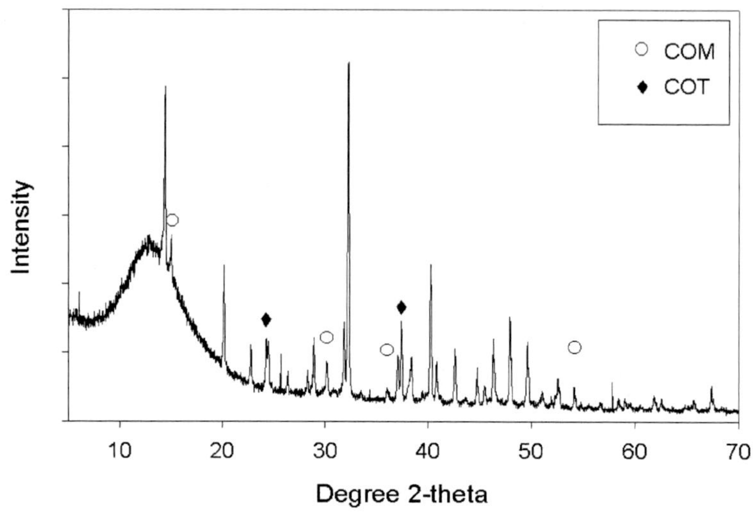

Fig. 4.3.6. Typical XRD spectrum of CaOx/SiO_2 composites formed in 10-40% sugar solutions at an initial CaOx conc. of 130 ppm

Fig. 4.3.7. Scanning electron micrographs of CaOx formed at an initial CaOx conc. of 130 ppm, initial pH of 8.0 and 60°C in sugar solutions of (A) 10%, (B) 25% and (C) 40%; (white circle denotes COD crystal)

The effects of initial COM, SiO_2 and sugar concentrations on the nature and morphology of the phases obtained from the binary-foulant sugar solutions are shown in Figs 4.3.8 and 4.3.9.

In general, COD was the major crystalline phase formed at an initial CaOx concentration of 130 ppm (Figs 4.3.8A-F) with small amounts of co-existing COM and COT crystals. An increase in the sugar concentration generally yielded smaller and flattened COD crystals but had little effect on the phase composition. Some of the COD crystals, especially those formed at higher sugar concentrations, showed rounded/corroded edges and contained agglomerates of micro-particles on their (011) and $(01\bar{1})$ faces (Figs 4.3.8A, 4.3.8D and 4.3.8F). COD crystals were also found to exhibit stellate shape (Figs 4.3.8B and 4.3.8F), a habit previously observed under high temperature and sugar concentration (Doherty et al., 1994). The crystal habits of COT, on the other hand, displayed triclinic platelets at low SiO_2 SS and a mixture of flattened platelets and needles at high SiO_2 SS (Figs 4.3.8A, 4.3.8D and 4.3.8E). For a binary foulant sugar solution of high SiO_2 SS (3.0) and low sugar concentration (10%) (Fig 4.3.8E), the quantity of COT crystals appeared to be higher than those obtained from other conditions.

Fig. 4.3.8. Scanning electron micrographs of composites formed in sugar solutions at an initial CaOx conc. of 130 ppm, initial pH of 8.0 and 60°C; at an initial SiO_2 SS of 1.8 in sugar solutions of: (A) 10%, (B) 25% and (C)-(D) 40%; at SiO_2 SS 3.0 in sugar solutions of: (E) 10% and (F) 25% (white circles denote COT crystal)

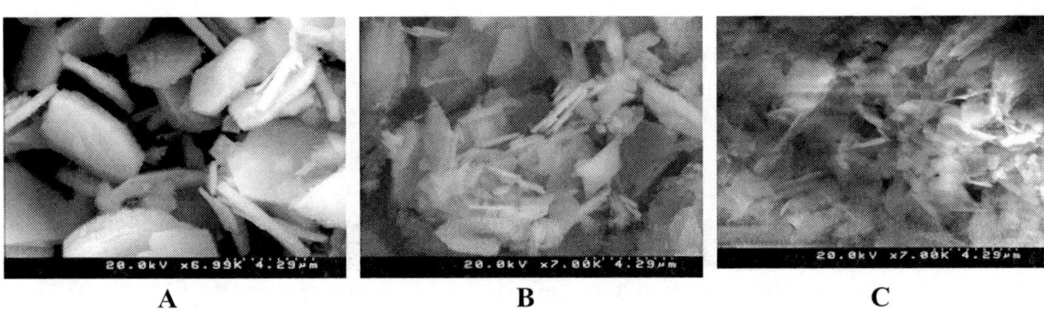

Fig. 4.3.9. SEM images of composites formed at an initial CaOx conc. of 44 ppm and SiO_2 SS 1.8 in sugar solutions of: (A) 10%, (B) 25% and (C) 40%

Interestingly, SEM imaging of composite materials obtained at an initial CaOx concentration of 44 ppm (Fig 4.3.9) showed a mixture of hexagonal platelets of COM and needles of COT. These phases were later confirmed by FT-IR analysis (XRD were not performed due to limited amounts of samples) and corresponded well with the CaOx solubility peak observed around 44 ppm in the previous batch study (section 4.2). Increasing the sugar concentration again showed little effect on the phase composition of the composites, although COM crystals were found to be smaller and were further flattened along the (100) direction to the extent that they became semi-transparent, a trend similar to that observed for COD crystals at higher CaOx concentration (Fig 4.3.8D) as well as COM crystals formed in CaOx-SiO_2 and CaOx-sugar precipitation systems (Figs 4.3.3C and 4.3.7C).

EDS quantitative analysis of the composite materials indicated the elemental distributions in the crystals. As an example, Fig 4.3.10 shows the average Si/Ca molar ratios at the surface and at the cross-section of the composite crystals obtained from sugar solutions. The analyses performed on the crystal cross-section revealed an average Si/Ca molar ratio of 0.2 and 0.1 for deposits obtained at CaOx of 44 and 130 ppm, respectively, lower than those (0.32 and 0.21) found at the crystal surfaces. Thus SiO_2 appeared to be concentrated mainly on the surface of the composites.

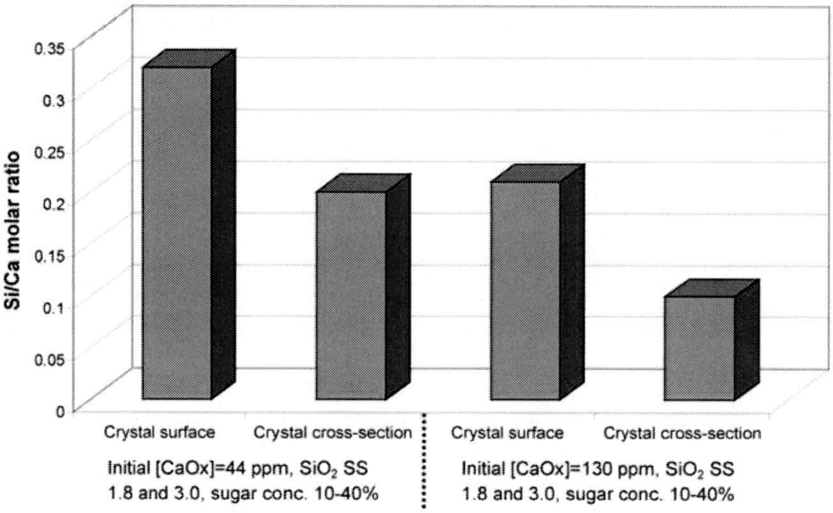

Fig. 4.3.10 Comparison of average Si/Ca molar ratios at crystal surface and cross-section of composites formed in sugar solutions at initial pH of 8.0 and 60°C

Figs 4.3.11A and 4.3.11B show that the CaOx precipitates obtained from 10 and 25% sugar solutions displayed IR spectra comparable to that observed in the absence of sugar, as anticipated from their XRD and SEM results. The spectrum of CaOx obtained in 40% sugar solution (Fig 4.3.11C) suggested the presence of COD with characteristic absorption peaks at 1643 cm^{-1}, 1327 cm^{-1} (C=O stretching) and 613 cm^{-1} (water libration). This was also evident by a loss of resolution in the 3000-3500cm^{-1} region, i.e., a strong peak at 3462 cm^{-1} (OH stretching vibration of H_2O) instead of five weak bands in the spectrum of COM (Fig 4.3.4A), and a reduced absorbance ratio between 781 cm^{-1} and 517 cm^{-1} peaks, probably caused by an increased number of hydrogen bonds involving O-C-O groups (Babic-Ivancic et al., 1985; Warpehoski et al., 1981). The typical IR spectra of the composites precipitated at 44 and 130 ppm CaOx (Fig 4.3.11D/E) showed major absorption peaks identical to the characteristic COM and COD bands, respectively, together with a weak band around 1108 cm^{-1} arising from the incorporation of amorphous silica.

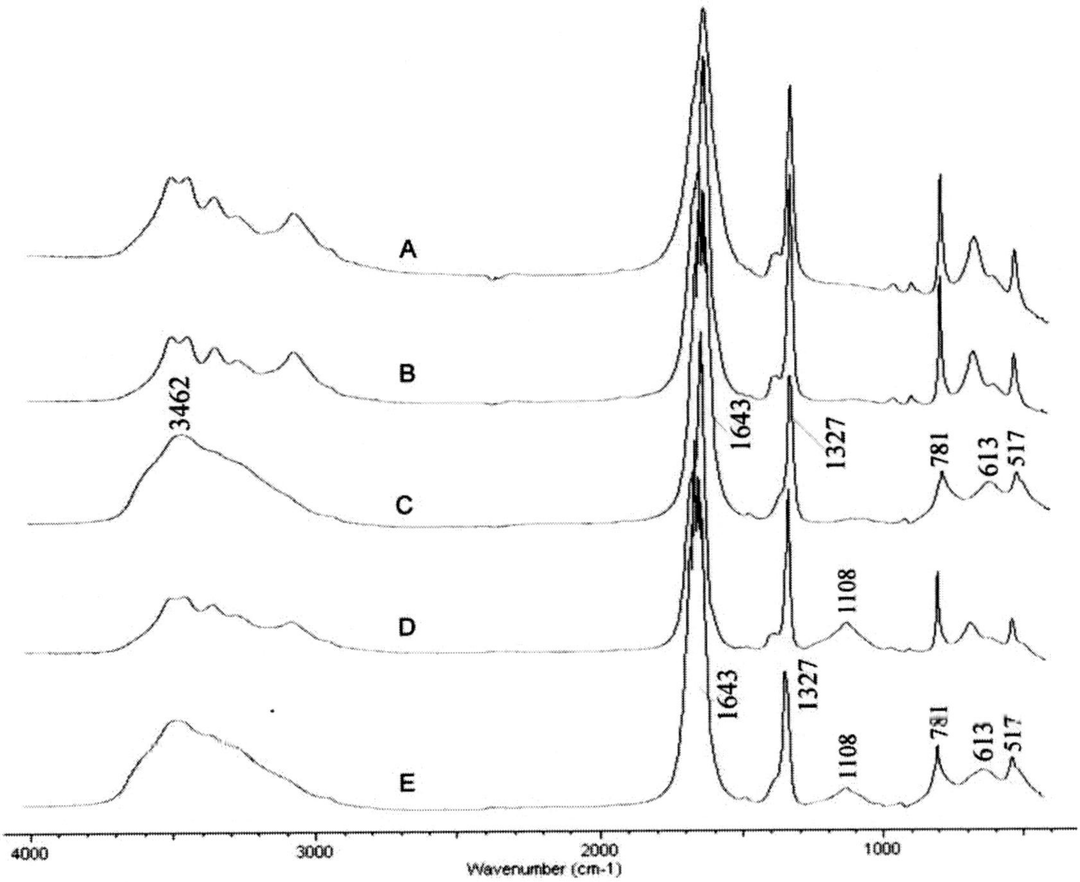

Fig. 4.3.11 FT-IR spectra of CaOx obtained at an initial CaOx conc. 130 ppm from sugar solutions of : (A) 10%, (B) 25% and (C) 40%; composites obtained from sugar solutions at an initial CaOx conc. of (D) 44 and (E) 130 ppm

Thus the characterization of the precipitates by various instrumental techniques provided valuable information on the crystallographic and morphological changes of CaOx and CaOx-SiO_2 composites in water and sugar solutions (see Table 4.3.1). The presence of SiO_2 or sugar in CaOx crystallization systems promoted the formation of COT and COD, respectively, whereas the presence of sugar in CaOx-SiO_2 mixtures gave rise to the stabilization of either COD or COT, depending on the initial concentration level of CaOx. The morphological impact of SiO_2 and sugar on CaOx was mainly to flatten and reduce the size of COM/COD crystals with increasing concentrations of the additives. These changes were expected to alter the kinetics and the thermodynamics of the co-precipitation behaviour of CaOx and SiO_2 in water and sugar solutions and would impact on the composite fouling mechanisms.

Table 4.3.1 Summary of crystallographic and morphological changes of CaOx and composites in water and sugar solutions

Sugar conc. (wt %)	Initial CaOx conc. (ppm)	Initial SiO_2 ppm/SS	Phase composition analysed from XRD	SEM analysis	
				Average crystal size (μm x μm) **	Major crystal morphologies
0	175	0	COM	(12-13)x4	Monoclinic prismatic
0	175	400/1.8	COM	(11-17)x5	as above
0	175	600/3.0	COT, minor COM	(<10)x(1-4)	Triclinic platelets and needles (COT); flattened hexagonal platelets (COM)
10	130	0	COM, little COD	(4-5)x2	Monoclinic (COM)
25	130	0	COM, minor COD	(2-3)x1.5	Monoclinic (COM); tetragonal with bipyramidal faces (COD);
40	130	0	COD, minor COM	(1.0-1.5)x (1.0-1.5)	As above
10	44	366/1.8	COM, COT *	(3.5-6.0)x(3-4)	Hexagonal platelets (COM); needles (COT)
10	130	366/1.8	COD, minor COT, COM and SiO_2	(2.0-3.6)x (2.0-3.6)	Tetragonal with bipyramidal faces and little corroded edges (COD); stellate (COM); triclinic platelets (COT)
10	130	600/3.0	as above	(1.8-3.6)x (1.8-3.6)	Tetragonal with bipyramidal faces and little corroded edges (COD); stellate (COM); triclinic platelets and needles (COT)
25	44	320/1.8	COM, COT *	(2.5-3.6)x1.5	Flattened hexagonal platelets (COM); needles (COT)
25	130	320/1.8	COD, minor COM, COT and SiO_2	(1.4-2.0)x (1.4-2.0)	Tetragonal with bipyramidal faces and little corroded edges (COD); stellate (COM); triclinic platelets (COT)
25	130	525/3.0	as above	(1.1-1.8)x (1.1-1.8)	Tetragonal with bipyramidal faces (COD); stellate (COM); triclinic platelets (COT)
40	44	250/1.8	COM, COT *	(2.2-2.5)x1.1	Flattened semi-transparent hexagonal platelets (COM); needles (COT)
40	130	250/1.8	COD, minor COM, COT and SiO_2	(0.8-1.4)x (0.8-1.4)	Flattened tetragonal with bipyramidal faces and corroded edges (COD); triclinic platelets (COT)

* determined based on FT-IR and SEM analysis due to limited sample quantities

** length x width

4.3.3 Discussion

It has generally been established that, in a supersaturated solution of CaOx where the precipitation of multiple crystal hydrates, i.e., mono-, di- and tri-hydrate is possible, the more soluble phases such as COD and COT usually form as precursors to COM formation, the least soluble and thermodynamically most stable phase, as a result of Oswald-Lussac rule of stages (Nancollas, 1982; Tomazic and Nancollas, 1980). The thermodynamically unstable phases appear rapidly in the initial stage of the reaction and *concomitantly* undergo phase transition through a dissolution – recrystallization process. Previous work (Manne et al., 1990; Nancollas, 1983; Sikiric et al., 1999) has suggested that this transformation process may be markedly influenced by the presence of inorganic or organic polymeric macromolecules in solution due to specific adsorption of these species at the solid-liquid interface as has been demonstrated for CaOx-SiO_2 composites (Fig 4.3.10). This could affect the rates of dissolution of the metastable phases (COT and COD) as well as the rates of nucleation and crystal growth of the stable phase (COM). In other words, the type of particular hydrate obtained will be reliant on kinetic rather than thermodynamic factors (i.e., different solubility products) (Garside et al., 1982).

Skiric et al. (1999) studied the phase transformation from COD to COM in various anionic and cationic surfactant solutions. It was shown that the rate of dissolution of COD was greatly reduced by the adsorption of surfactants on COD crystal faces. This in turn reduced the rate of supersaturation build-up necessary for COM crystallization. Other researchers (Cody and Cody, 1994; Manne et al., 1990) reported that polymeric additives at low polymer concentrations were capable of promoting the nucleation of COT from supersaturated CaOx solutions. Further increase in polymer concentration, however, would inhibit the growth of both COM and COT resulting in the precipitation of COD. Such transitions between CaOx hydrate forms in the presence of increasing additive concentration could be attributed to different adsorptive capabilities, Γ, of crystals of the above three hydrate forms for polymeric molecules, which followed a sequence of $\Gamma(COM) > \Gamma(COT) > \Gamma(COD)$ (Nancollas, 1982). Tomazic & Nancollas (Tomazic and Nancollas, 1980) studied the adsorption behaviour of the three CaOx hydrates with polymeric molecules. They found that the adsorption capacities varied considerably among the three hydrates, with COM adsorbing 3 times more than COT and 10-15 times more than COD. Therefore, small amounts of polyelectrolyes would be sufficient to retard the development of strongly absorbing COM nuclei to enable the formation of the least stable and moderate adsorbing COT, whilst higher polyelectrolye concentrations could further inhibit the growth of COT so that the growth of the least adsorbing COD crystals might proceed. The above adsorption sequence was supported by Figs 4.3.3C and 4.3.7C, which showed significantly greater inhibition of COM than that of COT, and in the case of COD, the nucleation was even enhanced. We could not rule out the possibility, however, that the latter occurred as a result of changes in the initial CaOx SS in solution (see sections 4.1 and 4.2). In addition, the absence of COD in the composites obtained at an initial CaOx concentration of 44 ppm in sugar solutions (Fig 4.3.9) was not expected according to this surface affinity sequence, which predicts stronger inhibition of both COM and COT than that of COD, and further experiments are necessary in order to clarify this.

Some insight into the nature of the differential adsorption affinities exhibited for the three CaOx hydrates could be gained by examining the crystal-polymer interactions based on the concept of molecular recognition (Mann, 1988; Weissbuch et al., 1991) associated with electrostatic accumulation and structural similarity. Recent work (Furedi-Milhofer et al., 2002) has suggested that the transformation of COM into COD in the CaOx-organic additive crystallization systems was strongly driven by variations in the ionic characters of the different crystal phases. The surface charge density of COD crystals may be shielded by a higher degree of surface hydration, which lowered the adsorbability of COD crystals due to reduced electrostatic interaction. On the other hand, structural fit between the lattice parameters of polymeric additives and ionic spacings on major COM crystal faces may be involved in the higher specific adsorption of organic macromolecules for this hydrate form (Furedi-Milhofer et al., 2002). Geometrical information on the crystal lattice of CaOx hydrates was not yet available, although similar structural correlation has been satisfactorily demonstrated for calcium phosphate precipitation systems in the presence of organic molecules (Furedi-Milhofer et al., 2002). Nonetheless, it follows from the above arguments that the formation of the stable COM phase might be retarded by a greater surface adsorption of polymeric SiO_2 and sucrose molecules due to higher surface charge and structural compatibility, which stabilises the metastable hydrates COD and COT. It is important to note that an increase in the COT crystal content in the presence of high SiO_2 SS (3.0) and low sugar concentration (10%) (Fig 4.3.8E) supports the general idea based on results from CaOx-SiO_2 and CaOx-sugar systems that the formation of COD and COT are *selectively* promoted by sugar and SiO_2 molecules, respectively. This, accordingly, may be explained by specific geometric relationship between active functional groups of SiO_2/sucrose molecules and interatomic distances on the crystal faces of COT/COD.

The main modification of crystal habits of COM and COD formed in water and sugar solutions, e.g., the flattening of crystals along certain faces, may be related to lattice matching, as mentioned earlier, and the stereochemical or orientational compatibilities of SiO_2/sucrose molecules and CaOx lattice ions at specific crystal planes. In COM structure, for instance, (010) faces are characterized by an alternation of oxalate ions lying perpendicular and parallel to the face, whereas a dense pattern of complexed calcium ions is exposed at (100) faces due to the oblique emergence of oxalate ions at the surface (Tunik et al., 1996). Consequently, specific adsorption may be induced by an optimal local co-ordination geometry as a result of stereochemical equivalence between the configuration of SiO_2/sugar functional groups (e.g., surface hydroxyl groups) and those of the anion (oxalate) motif of the crystal structure at (100) plane (Letellier et al., 1998; Mann, 1988).

The rounding and corrosion on the edges of some COD crystals along with the appearance of micro-crystals on their faces (Figs 4.3.8A/D/F) was clearly the result of COD $\rightarrow$ COM transformation in which the new phase developed on the surface of COD crystal during its dissolution. This observation demonstrated that the COD/COM phase transformation was via a secondary nucleation on surface rather than in the bulk solution.

The absence of COT in the IR spectra of samples, where COT was expected to be the major phase of CaOx according to XRD and SEM results, reflected a disadvantage of FT-IR analysis because of sample preparation procedure necessitating grinding and pelleting of the materials. This procedure may have led to the decomposition of the less stable form of CaOx hydrates due to heat generation (COT stability is highly temperature-dependent and the stability of CaOx hydrates follows a sequence of COM > COD > COT) (Deganello et al., 1981; Kontoyannis et al., 1997).

4.3.4 Conclusions

A combination of microscopic and spectroscopic analyses has been used to probe the crystallographic and morphological properties of CaOx formed in various media. The phase composition and morphology of CaOx were strongly dependent on the initial concentration of CaOx, SiO_2 and sugar. The presence of SiO_2 or sugar in CaOx precipitation systems selectively shifted the phase composition from COM to COT and COD, respectively, in a concentration–dependent manner with corresponding modification of the crystal size and habits of COM. CaOx-SiO_2 co-precipitation in sugar solution led to the stabilisation of either COD or COT at the expense of COM, depending on the initial CaOx concentration in solution. Increasing the concentrations of sugar in binary-foulant sugar solutions was found to result in smaller and flattened crystals of COM or COD with little impact on the phase composition of precipitates. The crystal habits of COT in the composites ranged from triclinic platelets to needle-like crystals according to the initial concentrations of CaOx and SiO_2 involved. The COD $\rightarrow$ COM transformation appeared to proceed via a dissolution and *surface* nucleation process. These control of crystal phases and growth habits of CaOx observed in water and sugar solutions has been attributed to preferential adsorption of SiO_2 and sucrose molecules based on electrostatic interactions as well as structural and stereochemical relationships.

Chapter 5

COMPOSITE FOULING OF CAOX-SIO_2 IN SUGAR MILL EVAPORATORS

5.1 FLOW BOILING HEAT TRANSFER OF WATER AND SUGAR SOLUTIONS

Flow boiling takes place in various industrial applications. In multi-effect sugar mill evaporators, juice concentration occurs via forced convective nucleate flow boiling in vertical calandrias. The performance of these evaporators, which accounts for ~70% of energy consumption in sugar mills, is important for efficient running of a sugar factory. Thus the ability to accurately predict the flow boiling heat transfer coefficients in sugar mill evaporators is desirable, not only for the optimal design and operation of equipment, but also as a means to help analyze and interpret the mechanisms involved in such heat transfer processes. This section describes an approach to correlate the heat transfer coefficients of sugar solutions during subcooled nucleate flow boiling in a vertical annulus based on modifications of existing models developed for pure water.

5.1.1 Previous Works of Flow Boiling Heat Transfer

Analysis of flow boiling heat transfer has been attempted by many workers (Celata et al., 1993; Chen and Tuzla, 1995; Muller-Steinhagen et al., 1986) using a widely-recognized Chen (1966) model, which was developed for saturated flow boiling but could be conveniently extended to subcooled boiling. The Chen model assumed that the flow boiling heat transfer coefficient was a modified sum of forced convection and nucleate boiling heat transfer coefficients. The convective and nucleate boiling terms were obtained by the Dittus-Boelter correlation (Winterton, 1998) and the Forster-Zuber (1955) pool boiling correlation, respectively. An enhancement factor, F, representing the ratio of the effective two-phase Reynolds number to the liquid Reynolds number, was included into the convective term to account for the enhanced convection due to bubble dynamics, whereas a suppression factor, S, was included into the nucleate boiling term to account for the reduced thermal boundary layer due to flowing conditions.

A number of variations on Chen's additive method have been tested for water and other organic fluids (Bjorge et al., 1982; Liu and Winterton, 1991; Steiner and Taborek, 1992). Liu and Winterton (1991), for example, used an asymptotic-type addition model for analyzing both saturated and subcooled flow boiling data. They replaced Forster-Zuber correlation with Cooper's pool boiling correlation (Cooper, 1984) for deriving the nucleate boiling term. The enhancement and suppression factors were also re-defined. Another Chen-type model was that described by Bjorge *et al.* (1982), which was an extension of the Bergles and Rohsenow (1964) correlation to cover subcooled and saturated regions utilizing pure water. The authors used a superposition equation similar to that described by Liu and Winterton except that an incipient boiling term suggested by Bergles and Rohsenow was included instead of the suppression factor. This resulted in good correlation for both saturated and subcooled boiling data.

Apart from the superposition models derived from mechanistic analysis [such as those of Chen (1966) and Rohsenow (1953)], much work has been published on empirical correlations for flow boiling heat transfer applications. Shah (1983) developed a model for the quick assessment of subcooled flow boiling heat transfer in annuli with a wide range of experimental parameters and a variety of fluids. The model divided the flow boiling curve beyond the single-phase region into partial and fully developed regions, where a dimensionless term, ψ_o, was introduced to represent the ratio between two-phase boiling heat transfer coefficient and the liquid only coefficient as a function of the Boiling number, Bo (which characterises the intensity of nucleate boiling). The expressions of ψ_o varied depending on the subcooled boiling region.

A dimensionless group model was developed by Moles and Shaw (1972) to correlate boiling heat transfer to subcooled liquids under forced convection based on the data from various heating surfaces and geometries. Dimensionless terms such as the Boiling number, Jacob number and Prandtl number were incorporated into the model to account for the effects of heat flux, liquid velocity and sub-cooling. The pressure effect was considered by the density ratio. Although models of this type provided a simple and comprehensive approach for evaluating and predicting the boiling heat transfer coefficients, their validity is often limited to the range of fluids and conditions used to formulate the correlations (Butterworth and Shock, 1982).

Most boiling heat transfer studies of sugar solutions reported so far have been based on pool boiling (Garyazha and Kulinchenko, 1975; Mayinger and Hollborn, 1977; Quraishi and Varshney, 1975) and little work [except for those of Watson (1988) and Starzak (1996), etc.] has been conducted to apply the existing models, whether analytical or empirical, to the estimation of boiling heat transfer coefficient data relevant to the multiple effect evaporators in sugar mills. This section, therefore, characterized the nucleate flow boiling heat transfer performance of water and sugar solutions in an annular arrangement under thermal hydraulic conditions relevant to those in the sugar mill evaporators. In particular, the effects of operating conditions including fluid velocity, heat flux, subcooling and sugar concentration on heat transfer coefficients were evaluated. The experimental data were compared with selected models (see Appendix IIIA) and a modification of Chen model was proposed which provided a better fit to measured heat transfer coefficients for both water and sugar solutions.

5.1.2 Flow Boiling Heat Transfer Data

5.1.2.1 Subcooled flow boiling regions

The regime of subcooled nucleate boiling is known to consist of two sub-regions (Fig 5.1.1A) depending on the levels of heat flux and wall superheat involved (Dhir, 1998; Prodanovic et al., 2002):

- partial or highly subcooled boiling region, and
- fully developed or low subcooled boiling region

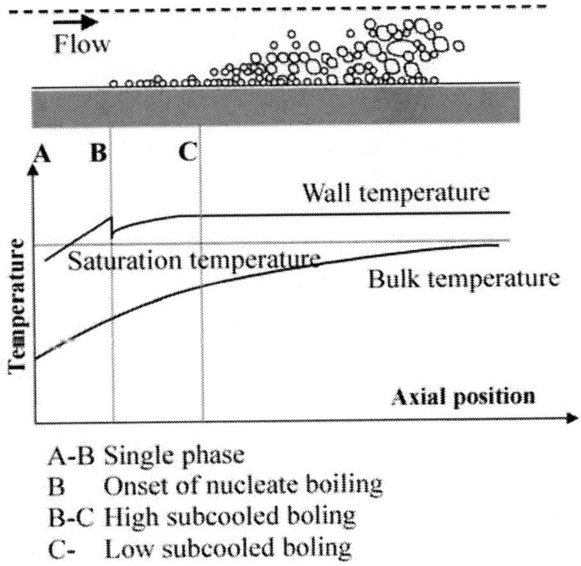

Fig. 5.1.1A Regions of subcooled boiling heat transfer (Prodanovic et al., 2002)

The former region starts at low to moderate wall superheat and heat flux values, where heat transfer is controlled by both forced convection and surface nucleate boiling since only a limited number of bubble nucleation sites are available on the boiling surface. The latter occurs at higher heat flux levels, where the contribution of forced convection diminishes with an increase in the number of active nucleation sites. Thus the liquid velocity and subcooling are generally found to affect the heat transfer coefficient in the partial boiling region but become negligible in the fully developed boiling region, where the heat transfer coefficient is controlled primarily by heat flux (Shah, 1983; Sivagnanam et al., 1994).

The method used to determine the region of subcooled boiling data in this work (Fig 5.1.1B) was that proposed by Shah (1983), which distinguishes different subcooled boiling regions based on the ratio between subcooling and wall superheat, $\Delta T_{sc} / \Delta T_{sat}$, with the transition point characterized by the following equations:

$$\Delta T_{sc} / \Delta T_{sat} = 2 \text{ (Bo} \geq \sim 2.5\text{x}10^{-4}) \tag{5.1.1}$$

$$\Delta T_{sc} / \Delta T_{sat} = 6.3 \times 10^{4} Bo^{1.25} \text{ (Bo} < \sim 2.5\text{x}10^{-4}) \tag{5.1.2}$$

i.e., the region is high subcooling if $\Delta T_{sc} / \Delta T_{sat} > 2$ or $\Delta T_{sc} / \Delta T_{sat} < 2$ and $\Delta T_{sc} / \Delta T_{sat} \geq 6.3 \times 10^4 Bo^{1.25}$, otherwise the region is low subcooling.

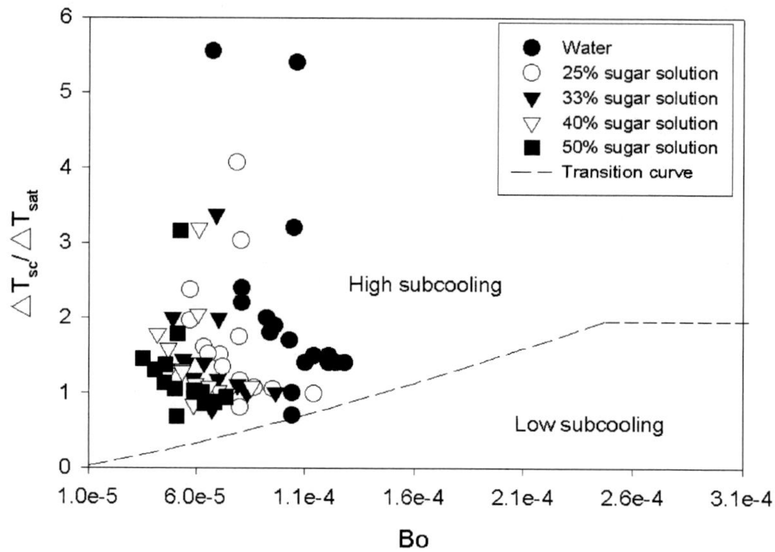

Fig 5.1.1B Identification of subcooled boiling regions for water and sugar solutions based on subcooling

As shown in Fig 5.1.1B, all boiling data lay within the high subcooling or partial boiling region under the current experimental conditions, indicating that both the thermal hydraulic conditions of fluid, such as velocity and subcooling, and heat flux are important factors in determining the flow boiling heat transfer coefficients of water and sugar solutions.

5.1.2.2 Effects of operating parameters

The variations of the heat transfer coefficients as a function of fluid velocity and subcooling were evaluated at comparable superheat levels for water and sugar solutions (Figs 5.1.2 and 5.1.3). Fig 5.1.2 shows that the heat transfer coefficient of water and sugar solutions increased with increasing fluid velocity (Fig 5.1.2) and could be described by a power law relation:

$$h \sim (v)^n \quad (5.1.3)$$

The slope n was found to be 0.49 for water and reduced from 0.38 to 0.17 as the sugar concentration increased from 25% to 50% (Table 5.1.1). A value of 0.8 would be expected if the heat transfer were in the non-boiling region, as indicated by the Dittus-Boelter equation (Wenzel et al., 1994). Thus the reduced dependence of the heat transfer coefficient on fluid velocity with increasing sugar concentration signified an increased contribution from the nucleate boiling term of heat transfer, which might be associated with reduced forced convective heat transfer due to lower Reynolds numbers of sugar solutions (Bao et al., 2000).

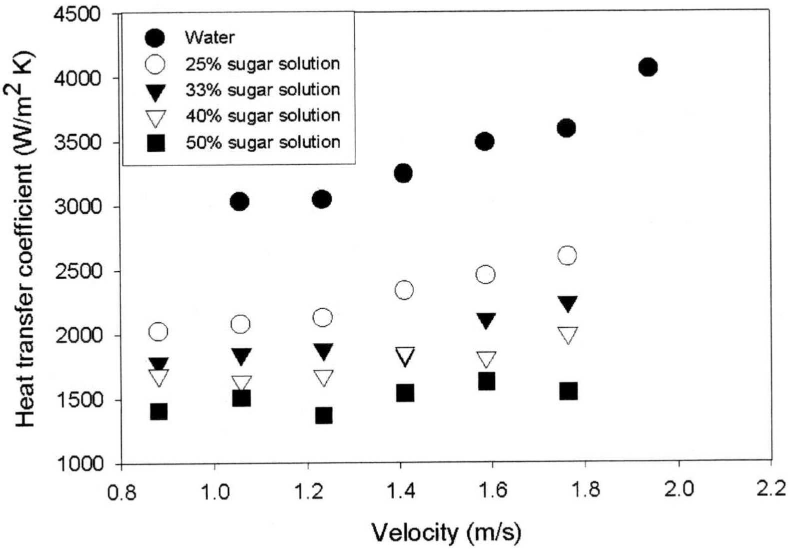

Fig 5.1.2 The effect of fluid velocity on heat transfer coefficients of water and sugar solutions at 20 K subcooling and ~16 K superheat

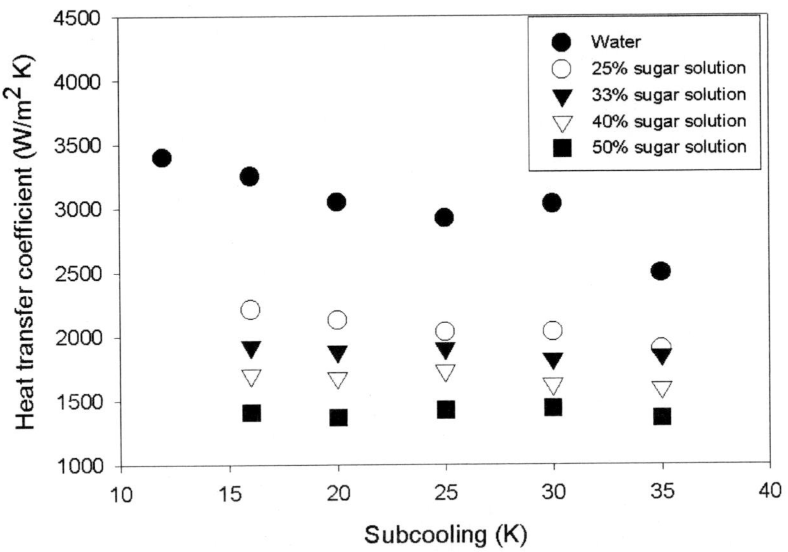

Fig 5.1.3 The effect of subcooling on heat transfer coefficients of water and sugar solutions at 1.2 m/s velocity and ~14 K superheat

Figure 5.1.3 shows that the heat transfer coefficients of water and sugar solutions decreased with increasing fluid subcooling. This was consistent with results obtained in previous studies using water and various organic mixtures (Muller-Steinhagen et al., 1986; Sivagnanam et al., 1994; Sivagnanam and Varma, 1990; Thom et al., 1965; Wenzel et al., 1994). The variations in the dependence of heat transfer coefficient on sub-cooling showed behaviour similar to that observed for fluid velocity, with the effect of subcooling being less pronounced in more concentrated sugar solutions (Fig 5.1.3).

Table 5.1.1 The dependence of heat transfer coefficient on fluid velocity and heat flux for water and sugar solutions

Condition	Exponent	
	$h \sim (v)^n$	$h \sim (q^{"})^m$
Water	0.49	0.60
25% sugar solution	0.38	0.60
33% sugar solution	0.32	0.69
40% sugar solution	0.25	0.61
50% sugar solution	0.17	0.72

Figure 5.1.4 depicts the effect of heat flux on the heat transfer coefficients of water and sugar solutions. For a given fluid velocity and subcooling, a general increase in the heat transfer coefficient with increasing heat flux was observed in all solutions tested, as expected for nucleate boiling regime (Thom et al., 1965). Similarly, the extent of heat flux effect could be represented using the following relationship:

$$h \sim (q^{"})^m \tag{5.1.4}$$

It has been reported in the literature (Wenzel et al., 1994) that exponent (m) values of 0 and ~0.8 are commonly distinguishable for forced convection controlled and nucleate boiling controlled regions, respectively. The values of m obtained in this work for water and sugar solutions were found to be between 0.60 and 0.72 (Table 5.1.1). These are consistent with the findings that all data were extracted from the partial boiling region. An increase in the effect of heat flux could also be noticed with increasing sugar concentrations, which again constituted a transition to more nucleate boiling heat transfer controlled region.

For a constant heat flux, the heat transfer coefficients of sugar solutions were considerably lower than that of water and decreased with increasing sugar concentration (Fig 5.1.4). The reduction in the heat transfer coefficients with increasing sugar concentrations has been reported in the previous studies (Cordiner, 1969; Mayinger and Hollborn, 1977), and could be attributed to several factors including differences in the physical properties, such as viscosity and wetting characteristics between water and the sugar solution, and the mixture effect, which arised from preferential evaporation of the lighter component (i.e., water) from a multi-component mixture (i.e., sugar solution) at the liquid-vapor interface (e.g., the bubble interface). The high viscosities of sugar solutions, particularly at the vapor-liquid interfacial layer, may impact on the heat transfer coefficient by suppressing the bubble growth rate and bubble-induced turbulence as well as diminishing heat transport by the liquid due to an additional mass transfer resistance in solution (Mayinger and Hollborn, 1977; Sivagnanam and Varma, 1990). Also, previous studies (Butterworth and Shock, 1982; Fujita, 1992; Wen and Wang, 2002) have shown that an increase in the wettability of a fluid could affect the heat transfer coefficient during nucleate boiling through its effect on the active nucleation site density on the surface. The more wettable fluids, which have smaller contact angles, as has been observed with increasing sugar concentration in solution (Fig 5.1.5), are more likely to "snuff out" cavities on heating surface thereby reducing the number of active nucleation sites available and heat transfer coefficient (Butterworth and Shock, 1982).

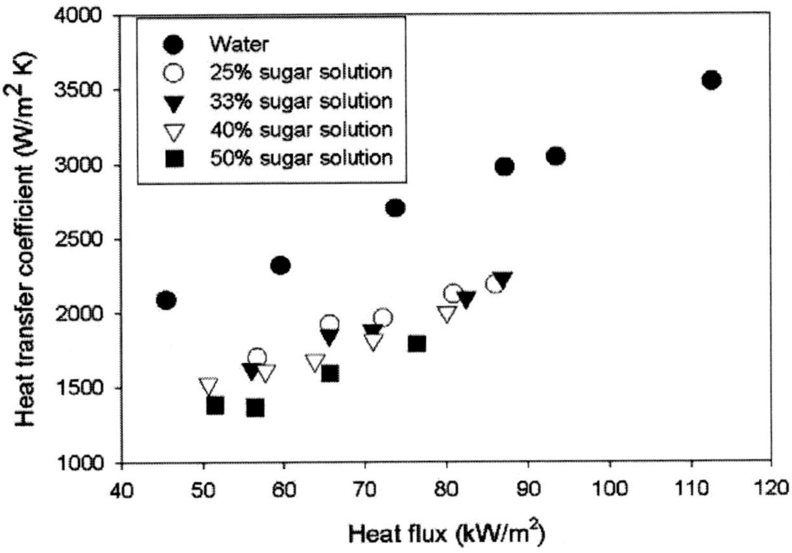

Fig 5.1.4 The effect of heat flux on the heat transfer coefficients of water and sugar solutions at 20K subcooling and 1.2m/s velocity

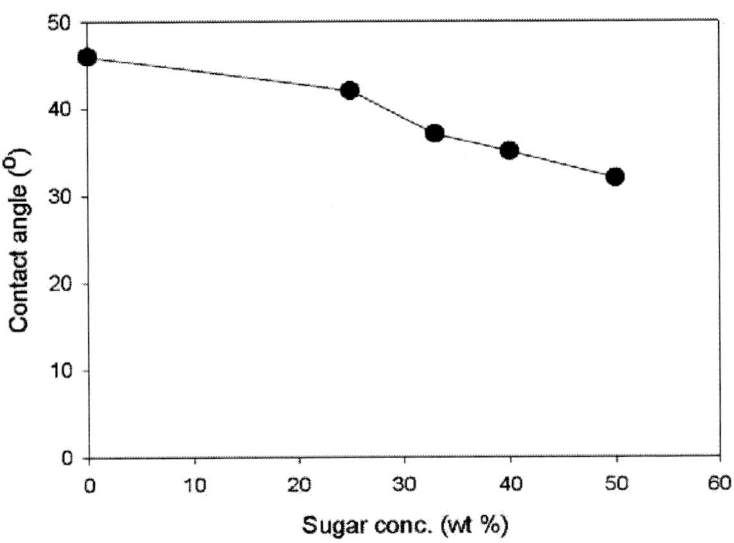

Fig 5.1.5 Variations of the static contact angles of water and sugar solutions on stainless steel tube as a function of sugar concentration.

Furthermore, the loss of solvent at the vapor-liquid interface resulted in a concentration gradient of solute (i.e., sugar) close to the interface, which drove a counter-diffusion of water molecules to minimize the sugar concentration at the interface as well as a local rise in the saturation temperature. This led to a decrease in the available wall superheat and hence boiling heat transfer coefficient in sugar solutions (Vadekar and Hills, 2001).

5.1.3 Data Correlation

The experimental data were correlated with a number of subcooled flow boiling models available in the literature, together with a new model based on a modification of the Chen model (see Appendix IIIA), which has been widely used in the design of heat exchangers in the process industry (Butterworth and Shock, 1982). Various modified versions of the Chen model have previously been proposed in an effort to improve its predictive ability for the different heating surfaces and fluids of interest. Peacock and Starzak (1996) developed a boiling model for climbing film evaporator system in a sugar mill using the Chen correlation, and found improved model performance when a Forster-Zuber constant was correlated with heat flux. Watson (1987) used the McNelly correlation (for nucleate boiling) in the Chem model to predict the heat transfer coefficient of Australian sugar mill evaporators. The calculated values were shown to match typical industrial values to within 30%. Bao et al.(2000) also obtained a better correlation of the data for organic fluids when the Forster-Zuber equation was replaced with the Cooper correlation. In our current work, the Chen correlation was modified by deriving the nucleate boiling components using the Rohsenow (1951) boiling correlation. This correlation was developed based on the transient conduction mechanism and micro-convection due to subsequent replacement of the superheated liquid layer by the departing bubbles from the surface, expressed in the form of dimensionless groups:

$$\frac{\mathrm{Re}_B \,\mathrm{Pr}}{Nu_B} = \frac{c_p \Delta T_{sat}}{h_{fg}} \tag{5.1.5}$$

or in the final form:

$$\frac{\Delta T_{sat} c_p}{h_{fg}} = C_{SF}\left[\frac{q''}{\mu h_{fg}}\sqrt{\frac{\sigma}{g(\rho_l - \rho_v)}}\right]^{0.33}\left[\frac{c_p \mu}{k}\right]^n \tag{5.1.6}$$

where the exponent n = 1.0 for water and 1.7 for all other fluids (Rohsenow, 1972). The surface coefficient, C_{SF}, which is a function of particular heating surface-fluid combination, takes into account the effects of surface conditions, fluid properties (e.g., viscosity and wettability) and their interactions on nucleate boiling heat transfer (Benjamin and Balakrishnan, 1997). In our current study, the influence of fluid properties on the values of C_{SF} was represented by fitting a hyperbolic equation to data using non-linear regression as follows:

$$C_{SF} = \frac{0.04}{1+(C/14.75)^{2.02}} \tag{5.1.7}$$

where C is the sugar concentration. Many studies (Nishikawa and Fujita, 1976; Pioro, 1999; Vachon et al., 1968) have found that the C_{SF} values for water is generally higher than those for the organic/inorganic solutions, corresponding well with data obtained in the present work (Fig 5.1.6).

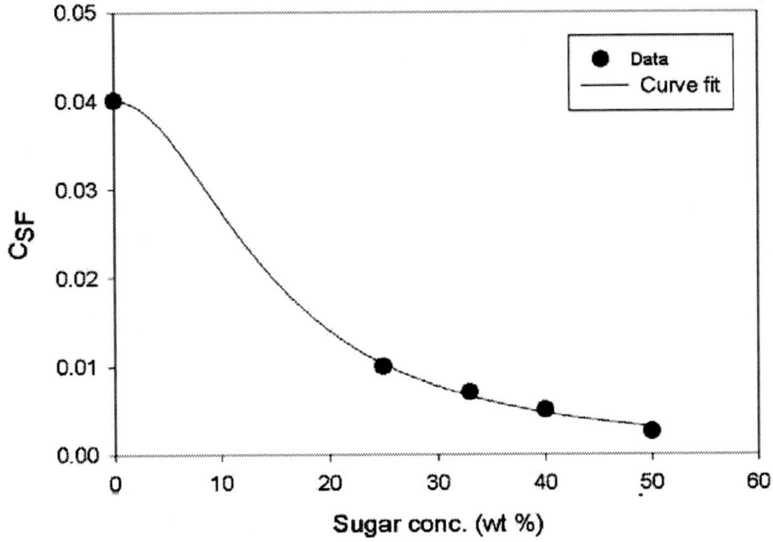

Fig 5.1.6 Variation of C_{SF} values as a function of sugar concentration

A revised Dittus-Boelter equation (eqn 5.1.8) was used in conjunction with the Rohsenow equation to correlate experimental data for water since it was found that those data did not compare well with the proposed model based on Chen's approach, even when the effect of C_{SF} was incorporated.

$$\frac{h_l D}{k} = 0.020\,\mathrm{Re}_l^{0.8}\,\mathrm{Pr}_l^{0.4} \tag{5.1.8}$$

This may be due to a greater contribution of forced convective heat transfer in water in comparison to those in the sugar solutions under the partial boiling condition (see Fig 5.1.2/eqn 5.1.3), and so an accurate prediction of heat transfer coefficient depended strongly on the proper choice of forced convection correlation. In fact, the use of the original Dittus-Boelter equation in Chen's correlation has always been questionable as the flow velocity distribution that normally occurs under the non-boiling conditions would often be disturbed by the bubble activities during nucleate boiling (Rohsenow, 1953).

Fig 5.1.7 shows the comparison of measured and calculated heat transfer coefficients for water and sugar solutions using the proposed model. Comparisons of all model fits to the data obtained in this work are summarised in Tables 5.1.2 and 5.1.3. In addition to modifying the Dittus-Boelter (D-B) equation, the Gnielinski correlation, which was shown to be more accurate than the D-B equation (Celata et al., 1993; Chen and Tuzla, 1995; Muller-Steinhagen et al., 1986), was also tested with slightly better data fits (mean deviation ~10% for water) than those of the D-B equation. This was, however, considerably worse than the data fits by the modified D-B equation proposed. Thus in this work the proposed model with modified D-B equation has been used to account for flow development under nucleate boiling.

Overall, the model proposed in this work fitted the data better than the other models tested, with a mean deviation of 3.8% and over 90% of the data points having less than 10% deviation (Table 5.1.2 and Fig 5.1.7). This was not surprising since the prediction of nucleate boiling coefficient by the other models did not account for the mixture effect and variations in the fluid wetting characteristics due to changes in sugar concentrations, which was shown to significantly affect nucleate boiling (Figs 5.1.2-4). The above effects were lumped into one surface coefficient C_{SF} in this study since the relative contribution of each of these phenomena in nucleate boiling was still not determined; also, the contact angles of water and sugar solutions (Fig 5.1.5) could not be corrected for equivalent flow boiling conditions due to the lack of temperature coefficient data. As a result, the proposed model was only applicable to water and sugar solution. For further improvement of the model and its generalization to other media, separate terms for viscosity, wettability and mixture effects may be considered.

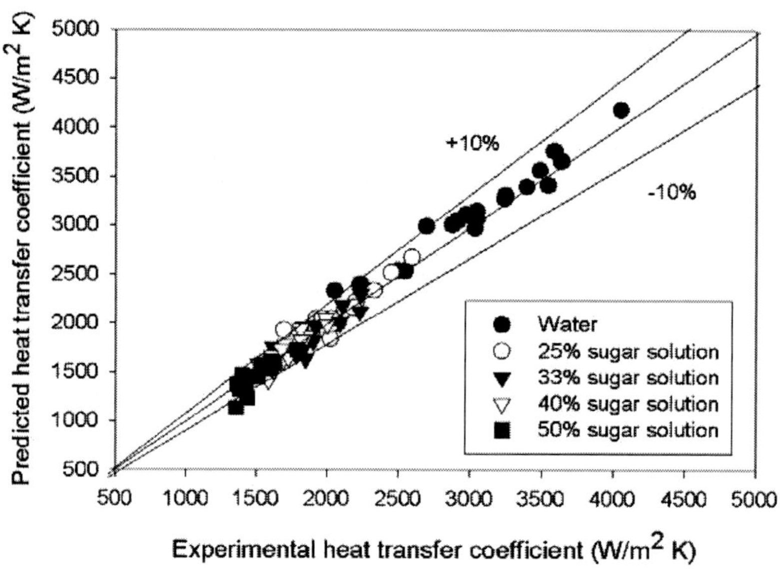

Fig 5.1.7 Comparing measured and calculated heat transfer coefficients for water and sugar solutions using the proposed model based on Chen's approach

The Liu & Winterton model and the Bjorge et al. model gave data fits only slightly lower than that obtained for the proposed model (mean deviation ~7%). A detailed analysis, however, indicated that the data fits by the Liu & Winterton model and the Bjorge *et al.* model were rather unevenly distributed over different sugar concentrations and appeared to deteriorate at either the lowest or the highest limit of sugar concentrations (i.e., for water or 50% sugar solution, see Table 5.1.3). The lower model fits for water by these two correlations, as noted above, might be the result of inaccurate prediction of forced convection coefficient by the Dittus-Boelter equation, and errors associated with values of fluid molecular weight in the Liu & Winterton model (see Appendix IIIA, eqn A10).

Table 5.1.2 Comparison of model fits to subcooled flow boiling data for water and sugar solutions

Model	Mean. dev * (%)	Avg dev ** (%)	Max dev (%)	Calc. data % ≤ ±10% dev range
Chen (1996)	21.9	21.8	41.4	9.5
Liu & Winterton (1991)	7.3	0.7	19.7	70.2
Moles & Shaw (1972)	39.6	-39.6	60.7	0
Shah (1983)	17.8	17.6	29.5	13.1
Bjorge et al. (1982)	7.1	-1.7	19.2	77.4
Proposed	3.8	0.2	16.7	91.7

* Mean deviation is calculated as: $1/n\sum_{n}\left|\left[h_{cal}-h_{\exp}\right]/h_{\exp}\right|\times 100\%$

** Average deviation is calculated as: $1/n\sum_{n}\left[h_{cal}-h_{\exp}\right]/h_{\exp}\times 100\%$, where n is the number of data

Table 5.1.3 Comparison of mean and average deviations in fitting the experimental data to var1ious models

Model	Water	Sugar solutions 25%	33%	40%	50%
Chen (1966)	15.6/15.5	23.2/23.2	23.4/23.4	23.9/23.9	24.8/24.8
Liu & Winterton (1991)	12.3/12.2	4.0/2.8	4.5/-1.2	4.8/-3.7	9.6/-9.6
Moles & Shaw (1972)	37.6/36.6	38.8/38.3	38.9/39.6	41.2/40.8	43.5/43.5
Shah (1983)	17.7/17.6	22.9/22.9	20.2/20.2	17.6/17.6	10.6/10.1
Bjorge et al. (1982)	8.3/5.2	5.2/3.3	5.2/-1.5	6.3/-4.7	13.0/-12.8
Proposed	3.7/3.1*	3.2/1.0	4.1/-1.2	3.9/0.5	4.2/-3.3

* mean deviation/average deviation

The Chen model and the Shah model were found to over-predict the flow boiling heat transfer coefficient by an average of ~20%, considerably higher than those achieved for three afore-mentioned models. The poor performance by the Chen model was consistent with results obtained previously (Campbell et al., 2000; Kandlikar, 1983; Prodanovic et al., 2002), which showed that the Chen model typically over-predicted the effect of nucleation resulting in large deviation. On the other hand, Prodanovic et al. (2002) reported that the Shah model exhibited good agreement with data in the fully developed boiling region but tended to over-predict the heat transfer rate in the partial boiling region. The Moles & Shaw model showed the worst data fit with under-predictions by as much as 61%, demonstrating the limited range of applicability characteristic of the dimensionless group correlations (Guglielmini et al., 1988).

Thus the model proposed in this work based on Chen's approach provided the best estimate of subcooled nucleate flow boiling heat transfer coefficients in sugar solutions for a range of thermal hydraulic parameters and may be used to qualitatively predict the heat transfer performance of sugar mill evaporators under various operating conditions. Although the efforts of modifying the existing flow boiling correlations are by no means exhaustive, they certainly open perspective to improve the design and operation of the evaporator system of sugar mill using well-established methods.

5.1.4 Conclusions

The boiling heat transfer characteristics of water and sugar solutions have been examined in an annulus under subcooled nucleate flow boiling conditions. The flow boiling data were shown to be in the partial subcooled boiling regime, where forced convection and nucleate boiling occur simultaneously. As a result, the flow boiling heat transfer coefficient was affected by the thermo-hydraulic conditions of fluid, such as velocity and sub-cooling, as well as heat flux. Increasing the sugar concentration resulted in a gradual increase in the contribution of nucleate boiling heat transfer and a decrease in the heat transfer coefficient. This could be ascribed to variations in the fluid properties including viscosity and wettability and a local rise in the saturation temperature of fluid at vapor-liquid interface due to a mixture effect.

The measured experimental data were successfully correlated to a subcooled nucleate flow boiling model developed based on Chen's approach. This new model incorporated the Rohsenow pool boiling correlation and allowed for the variation of a surface coefficient, C_{SF}, as a function of sugar concentration. The D-B equation was also revised to yield a better fit to boiling data for water. The new proposed model compared favourably with existing models extracted from the literature under the current experimental conditions and may assist future design and optimisation of sugar mill evaporator units. It was demonstrated that the predictive ability of the proposed model was restricted to water and sugar media but may potentially be improved by explicitly incorporating viscosity, wettability and mixture effects for use in other fluid systems.

5.2 Composite Fouling of COM/SiO_2 from Aqueous Solutions in Various Effects of Evaporators

This section presents a simple but novel experimental approach to investigating the composite fouling behaviour of COM and SiO_2 from water in a dynamic closed-loop system to gain further insight into the mechanisms governing the composite fouling of COM and SiO_2 and to help develop an effective predictive model. This approach allowed the simulation of the effect of feed concentration of successive stages of evaporator system within one experiment. The dynamic system consists of a circulating fouling loop under forced convective and nucleate flow boiling heat transfer and evaporation resulting in a gradual increase in the feed concentration. The experiments were conducted in aqueous solutions of COM or SiO_2 alone and their binary mixtures, to allow for comparative studies. Characterization of deposits obtained from the dynamic closed-loop tests was also carried out, as information on the physico-chemical properties of scale is considered essential for better process control and the development of novel treatment procedures for scale control and removal.

5.2.1 Evaporator Fouling in Single Systems

5.2.1.1 COM Fouling

Figures 5.2.1a and 5.2.1b show the fouling curves of COM and the corresponding COM SS curves at COM initial concentrations of 20, 50 and 100 ppm (Appendix V). The onset of COM fouling occurred almost immediately after the experiments began (CF above 1.0). The instantaneous fouling resistance from an initial COM concentration of 50 ppm was found to be the highest at any given CF level. This is probably due to the formation and deposition of metastable colloidal COM particles in solution, which were not separated by filtration with 0.22 μm pore size membrane filter before ICP analysis and therefore contributed to a higher degree of COM supersaturation observed in solution (Table 5.2.1). This hypothesis was verified by passing COM solutions through a series of membrane filters (pore size: 2.5-0.005 μm) and analyzing the resulting COM concentrations after each filtration. For the other two runs, the instantaneous fouling resistance from an initial COM of 100 ppm was found to be lower than that from 20 ppm COM. Lencar and Watkinson (2001) also found that the initial fouling rate of COM in a double pipe cooling heat exchanger decreased as the initial relative supersaturation of COM in the feed increased from 1 to 14.5. These observations could be due to the high precipitation rate of COM in the bulk solution at high initial COM SS.

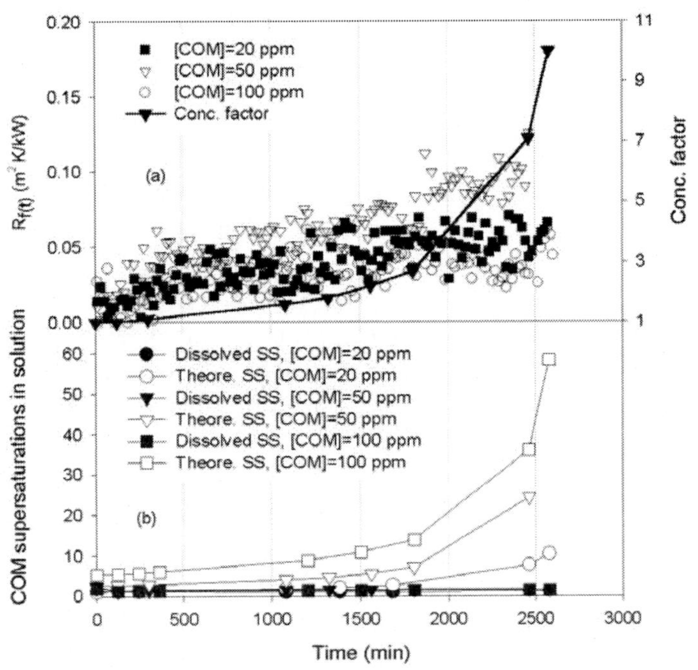

Fig. 5.2.1 COM fouling in single systems, (a) fouling resistance profiles at different initial COM concentrations, (b) Changes in COM SS in solution

The instantaneous fouling resistances for all the COM single systems obtained in this work were in the range 0.05 – 0.1 (m^2 K/kW) [The heat transfer resistance of a clean tube is ~0.56 (m^2 K/kW)]. These were thus much lower than those of Lencar and Watkinson and as a result very little COM scale was observed on the heat exchange surface. This difference may be explained by a lower range of initial COM supersaturation used in this study and the fact that our solutions were concentrated without further chemical additions, whereas Lencar and Watkinson (2001) used constant feed conditions by injecting makeup chemicals periodically without evaporation. Also, in this work bulk precipitation of COM in the feed tank was observed due to its normal solubility and high rate of precipitation under the existing operating conditions, in agreement with our previous batch tests (see section 4.1).

Table 5.2.1 Effect of supersaturation ratio on fouling of COM single systems

Initial COM conc. (ppm)	Initial supersaturation ratio	Equilibrium supersaturation in solution	* $R_{f(t)}$ at CF = 4.0 (m^2 K/kW)x10^2	* Linear fouling rate (m^2 K/kW min) x10^5
20	1.1	1.0	5.2	1.8
50	2.6	1.2	8.4	3.1
100	5.3	1.0	3.8	1.1

* Estimated from linear regression analysis

5.2.1.2 SiO_2 Fouling

The SiO_2 fouling resistance and the SiO_2 supersaturation curves are shown in Fig 5.2.2. The fouling resistance of SiO_2 single system remained almost negligible at low CF levels (1.0-3.0) before rising sharply when CF reached above 3.0 (theoretical silica SS ~5.3). This is apparently due to a drastic increase in the difference between total and theoretical SiO_2 supersaturations (supersaturations based on total and 'theoretical' silica concentrations, i.e., silica concentration in the absence of precipitation, respectively; also see Table 3.2) in solution from this theoretical SS level (Fig 5.2.2b), which represented the proportion of SiO_2 removed from solution (adhered to the surface and mostly precipitated in the tank) as evaporation continued.

Additional information on silica fouling was obtained by monitoring the changes in the dissolved SiO_2 supersaturation levels in solution during the experiment (Fig 5.2.2c). It was found that the dissolved SiO_2 supersaturation ratios in solution was mainly below 2.0 at CF 1.0-3.0 then markedly increased from 2.0 to 2.5 as the fouling resistance steeply rose to ~1.0 (m^2 K/kW). It may be inferred from these results that direct deposition of molecular silica (surface polymerization) may have occurred at low CF, which was then followed by particulate deposition of colloidal SiO_2 species resulting in the steep rise in the fouling resistance (Andritsos and Karabelas, 1999; Chan, 1988) (also see section 4.1). The above two fouling mechanisms are known to yield SiO_2 deposits of different physical characteristics: monomeric SiO_2 species forms an impervious film while colloidal silica particles produces a porous layer (Chan, 1988). This change may alter the thermophysical properties (e.g., thermal conductivity) of silica scale and could well explain the fouling resistance curve in Fig 5.2.2a.

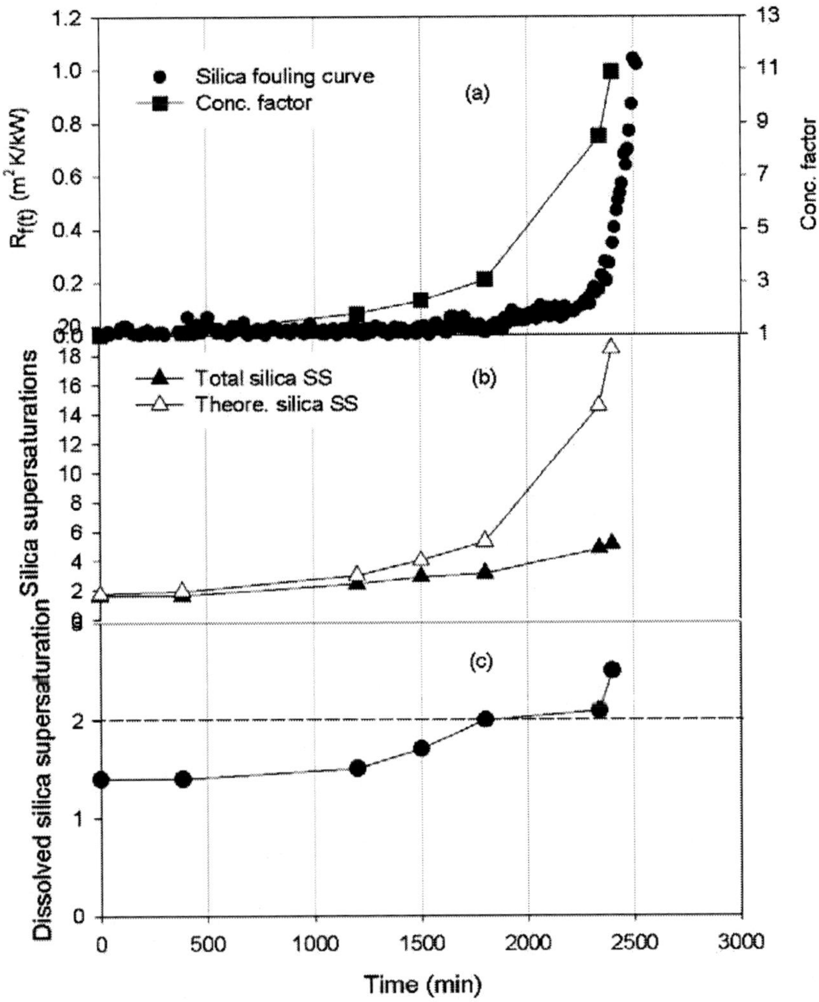

Fig. 5.2.2 SiO_2 fouling at initial conc. of 500 ppm

5.2.2 Evaporator Fouling in Binary Systems

5.2.2.1 Data Reproducibility

The reliability of the experimental data was evaluated by having a duplicate run for one of the binary systems (Fig 5.2.3). The results show that the fouling resistance data were reproducible with 90% of the data points having less than 10% deviation, a low magnitude considering the sources of errors expected.

5.2.2.2 Effect of COM/SiO_2 Ratio

Figs 5.2.4 – 5.2.5 present the data obtained for COM/SiO_2 composite fouling in the binary systems together with those of the corresponding single systems. The data for an initial COM concentration of 50 ppm (Fig 5.2.4) showed that composite fouling resistance started to rise at a lower degree of theoretical SiO_2 supersaturation (~2.0) (CF ≅ 1.4) than that observed for pure SiO_2 fouling (~5.3) (CF ≅ 3.0), indicating synergistic effect between the two scale components.

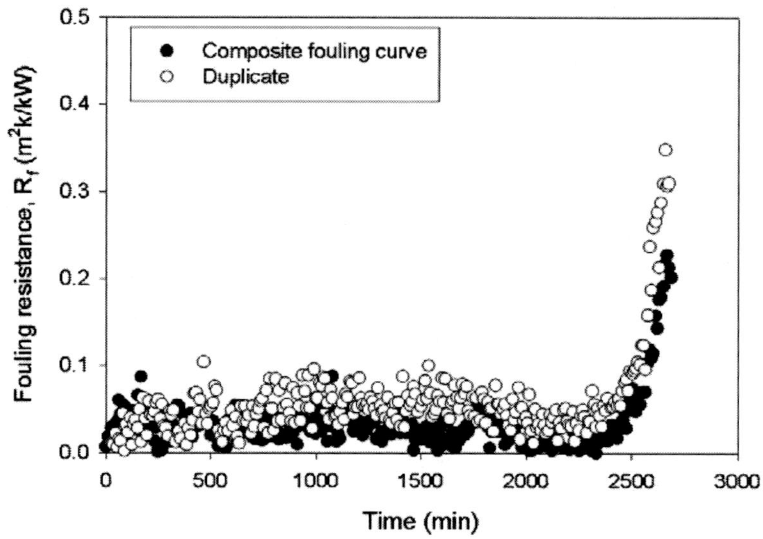

Fig 5.2.3 Reproducibility of dynamic experimental data

This phenomenon may be explained on the basis of the batch test results reported in section 4.1 in which the presence of COM enhanced the rate of SiO_2 polymerization. This in turn would increase the particulate deposition of colloidal SiO_2 species during the fouling process leading to a decrease in the degree of supersaturation required for composite fouling. However, such a synergistic behaviour was not observed for composite fouling resistances at initial COM concentrations of 20 and 100 ppm (Fig 5.2.5), both of which remained lower than those of the comparative single systems up to a theoretical SiO_2 SS of 9.0 (CF ~5.0) even though under those conditions SiO_2 was expected to have high polymerization rates as shown in the previous work of batch tests (section 4.1). Thus it is possible that other mechanisms may also occur during composite fouling of COM and SiO_2.

The effect of precipitating species on particulate fouling has previously been reported (McGarvey and Turner, 1999). The deposition rate of silt and hematite particles, for example, was enhanced in the presence of calcium carbonate precipitation (McGarvey and Turner, 1999). The increased deposition rate of silt may be accounted for by a reduced energy barrier between negatively charged silt particles and negatively charged stainless steel wall due to the incorporation of crystallizing calcium carbonate (calcite) into the silt particles. Similar interaction has also appeared in membrane fouling (Hong and Elimelech, 1997). Hong and Elimelech (1997) examined the fouling of nanofiltration membranes by natural organic matter (NOM). The rate of fouling was significantly increased in the presence of divalent cations such as calcium, owing to a decrease in the overall surface charge of NOM and the membrane. Thus, the high composite fouling rate observed for solution containing 50 ppm COM and 500 ppm SiO_2 in this study may be due to the lower interfacial energies between the surface of heat exchange tube and COM/SiO_2 colloidal particles as a result of charge-neutralization between positively charged COM particles and negatively charged SiO_2 particles (Callejas-Fernandez et al., 1991; Tadros and Lyklema, 1969).

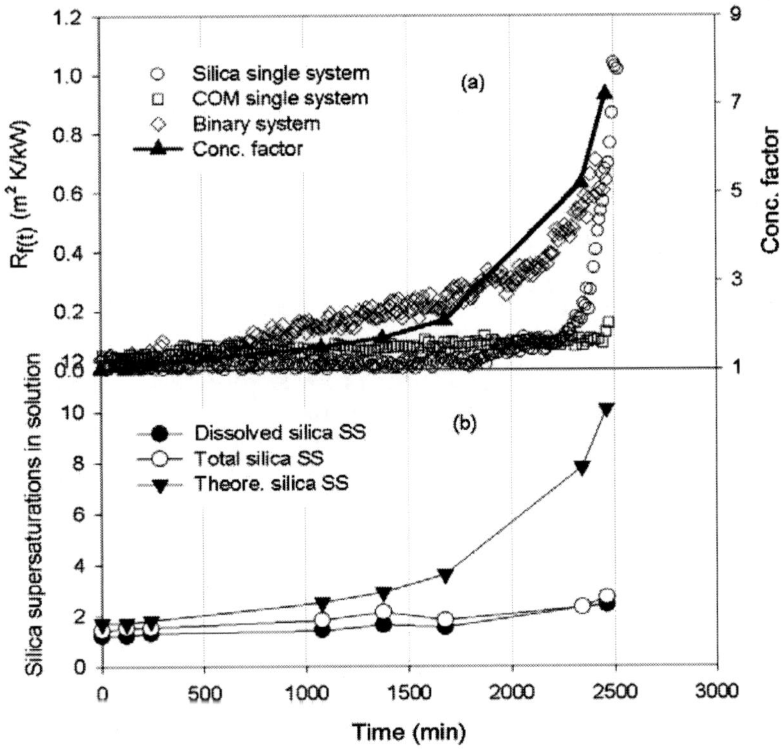

Fig. 5.2.4 Composite fouling of COM/SiO_2 at initial SiO_2 conc. of 500 ppm and initial COM conc. of 50 ppm.

Accordingly, the low fouling rates observed at initial COM concentrations of 20/100 ppm were probably due to the formation of colloidal COM/SiO_2 particles with surface charges that remained sufficiently high to prevent particle deposition. Moreover, the presence of high COM concentration (100 ppm) may have led to rapid aggregation and precipitation of COM/SiO_2 complexes in the bulk solution owing to the fact that the kinetics of COM precipitation increased as the initial COM supersaturation in solution increased (section 4.1).

On the other hand, the physical properties of fouling particles such as size and shape may also have an impact on particulate deposition as they determine particles' transport regime as well as their attachment efficiency to the wall (Epstein, 1988; Yiantsios and Karabelas, 1998). Hence, the different composite fouling rates of COM and SiO_2 observed in the binary systems may also be related to the variations in the size distribution of particles formed at different initial COM supersaturation levels (Furedi-Milhofer et al., 1990).

Thus the approach used and the dynamic test results described above demonstrated that the fouling resistance of COM started to rise at an earlier stage of evaporation cycle (theoretical COM SS $\geq$1.0) than that observed for SiO_2 fouling (theoretical SiO_2 SS $\geq$5.3), albeit to a much smaller extent than that of SiO_2 fouling resistance. The composite fouling resistance in the presence of 50 ppm COM started to rise from a theoretical SiO_2 SS level of ~2.0 corresponding to the 3rd effect of a sugar mill evaporator station, indicating that there is a synergistic effect of COM on composite scale formation.

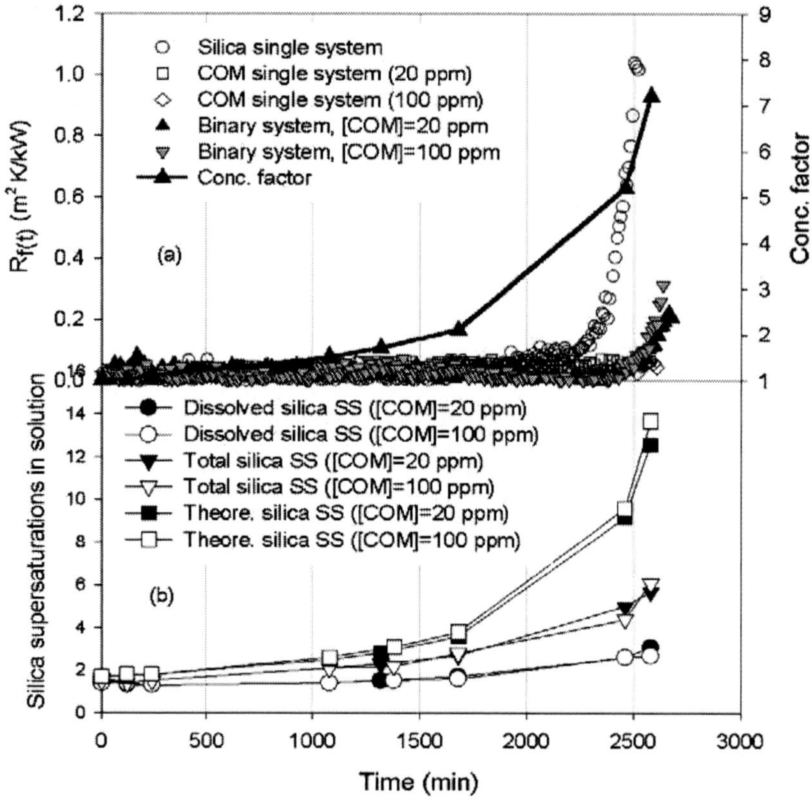

Fig. 5.2.5 Composite fouling of COM/SiO_2 at initial SiO_2 conc. of 500 ppm and initial COM conc. of 20 and 100 ppm.

5.2.3 Scale Deposits

5.2.3.1 Visual and Tactile Examination

The amounts of scale deposited onto the surface of the heat exchange tube were found to be small for all the COM single systems, in line with the low COM fouling resistances observed. Visual examination of the heat exchange surface during the later stage of SiO_2 fouling experiment found that silica scale had a rippled surface (Fig 5.2.6d) similar to that described for SiO_2 in another study (Bott and Gudmundsson, 1978). This rippled layer of deposits contained aggregates of large particles and was partially swept away upon draining the test section to remove the tube, uncovering a uniform inner layer of hard scale (Fig 5.2.6d). This inner layer adhered to the tube wall and could only be removed by mechanical means. These observations corresponded well with the proposition that the deposition of monomeric silica and colloidal silica species occurred at various stages of fouling.

In experiments with the composite scales, only the one obtained at an initial COM concentration of 50 ppm led to a complete coverage of tube surface (Fig 5.2.6b). As was obvious when scraping off the deposits, this scale was more tenacious than the inner layer scale formed in the SiO_2 single system. Composite deposits formed at initial COM concentrations of 20 and 100 ppm (Figs 5.2.6a and 5.2.6c) were composed of large isolated clusters, which appeared to be of similar strength to that of pure SiO_2 scale but less tenacious than the scale obtained with 50 ppm of COM.

(a) (b) (c) (d)

Fig 5.2.6 Scales from COM/SiO_2 binary systems with various initial COM conc.: (a) 20 ppm, (b) 50 ppm, and (c) 100 ppm and (d) SiO_2 single system

5.2.3.2 SEM-EDS Analysis

Scanning electron microscopy was used to evaluate the structural and morphological features of the scale deposits. Figs 5.2.7 and 5.2.8 present the electron micrographs of SiO_2 and composite scales at various magnifications.

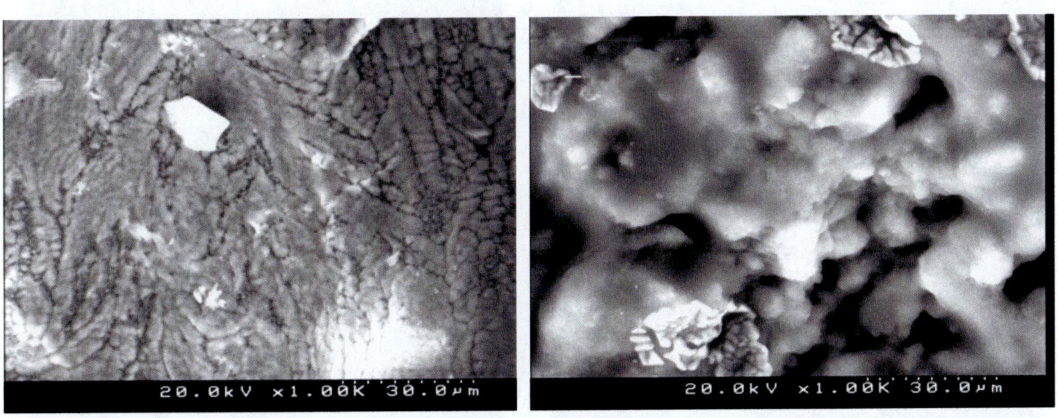

(a) (b)

Fig 5.2.7 Scanning electron micrographs of SiO_2 scales, (a) inner layer, (b) outer layer

It was evident from Figs 5.2.7a and 5.2.7b that the inner layer scale of SiO_2 has a different morphology from the outer layer of SiO_2 scale, and also from the three composite scales (see Fig 5.2.8). The inner layer scale appeared to be more compact than that of the outer layer scale, which appeared to have a porous structure. As mentioned earlier in fouling data analysis, this difference in the structure would be expected to result in a higher thermal resistance for the outer layer deposit, in comparison to the dense inner layer deposit under the flow-boiling conditions. The pores of the deposit would have been filled with a mixture of water and steam, which had a lower thermal conductivity than water (Turner et al., 2000).

The composite deposits (Fig 5.2.8a-c), on the other hand, exhibited a layered structure and a certain surface fragmentation indicating periodic growth of different fouling species on the solid phase and changes in the crystallinity of scales (Walthew and Turner, 1995).

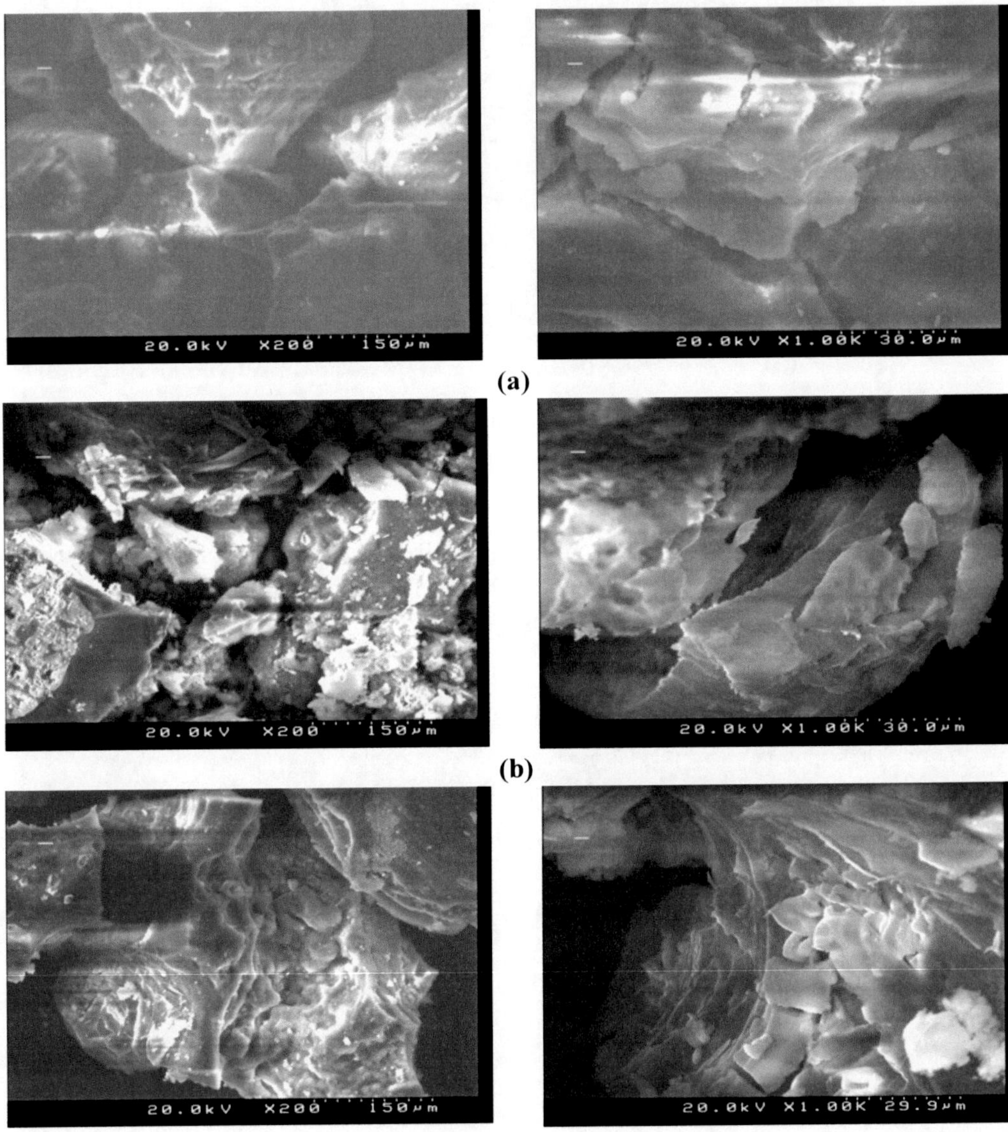

Fig 5.2.8 Scanning electron micrographs of composite scales, (a) COM 20 ppm, (b) COM 50 ppm, (c) COM 100 ppm

A closer look at the cross sections of the scales showed that the composite scale from 50 ppm COM (Fig 5.2.9b) was more densely packed in comparison to the SiO_2 inner layer scale, which appeared to contain a matrix of spheroids (Fig 5.2.9a). This difference might explain why this deposit was more tenacious than the SiO_2 scale. Previous studies (Bramson et al., 1995; Sudmalis and Sheikholeslami, 2000) on $CaSO_4$ and $CaCO_3$ fouling also found that the presence of $CaCO_3$ led to an increase in the strength of $CaSO_4$ scale. The authors suggested that the co-precipitated $CaCO_3$ might act as a bonding agent, cementing the structure of the $CaSO_4$ scale layer. It is likely that similar effect may have occurred in this study.

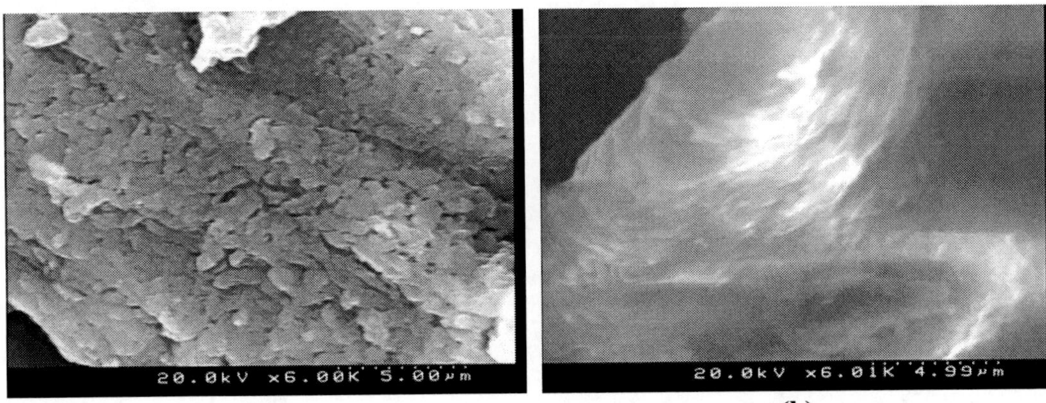

(a) **(b)**

Fig 5.2.9 SEM cross-sections of SiO_2 scale and composite scale, (a) pure SiO_2 scale and (b) composite scale at COM 50 ppm

Energy dispersive x-ray microanalysis (EDS) was used to identify the main elements in the scale samples. A typical elemental profile for the composite scale, shown in Fig 5.2.10a, revealed the presence of Ca in the deposit, which was later confirmed by XRD to be associated with COM (see next section) in addition to Si and O as the major components of both silica and composite scales (Fig 5.2.10a-b). Small levels of Na and Cl were also present, possibly because of the entrapment and/or absorption of NaCl from the solution.

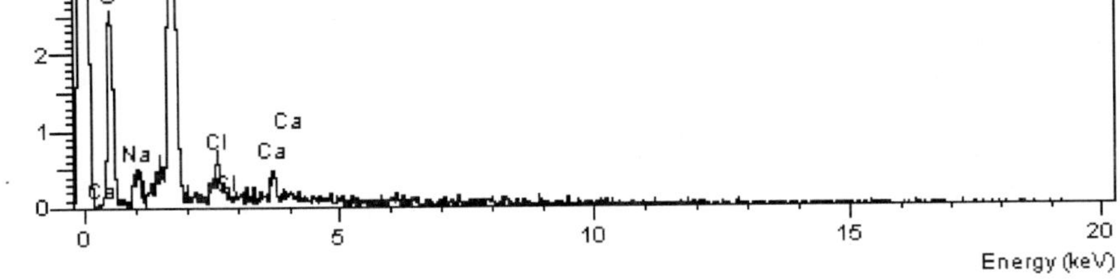

Fig 5.2.10a EDS spectrum for composite scale obtained at initial COM conc. of 100 ppm

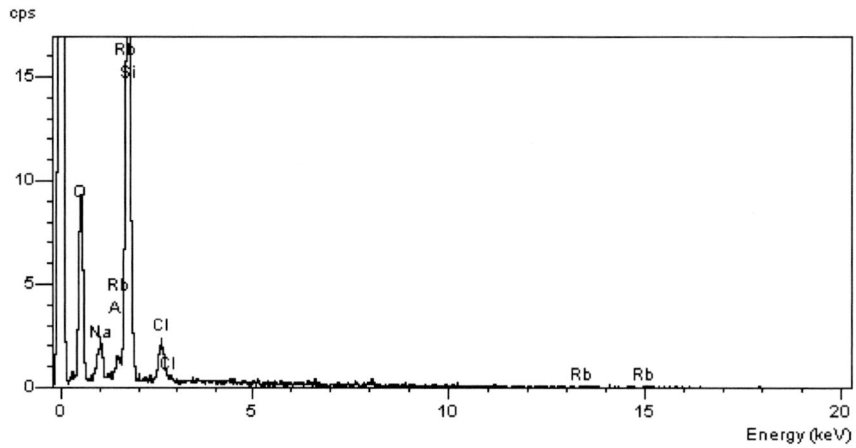

Fig 5.2.10b EDS spectrum for SiO_2 scale (inner layer)

To obtain further information on the composition of composite scales, samples were either treated with resin setting (to examine particle cross-section) or mixed with potassium bromide (KBr) powder and made into KBr pellets in a die (to examine particle surface) to give polished sections for x-ray analysis (see section 3.3). As indicated by high molar ratios between SiO_2 and COM at the cross-sections of the sample particles (Table 5.2.2), COM was located mainly in the interior of the particle while large amount of SiO_2 was present at the particle surface. The uneven distribution of COM might be explained by the initial formation of COM/SiO_2 complex or COM particles, which served as additional nucleation sites or growth center for subsequent SiO_2 polymerization and adsorption (see section 4.1).

5.2.3.3 XRD Analysis

X-ray powder diffraction analysis was conducted on the SiO_2 and composite scales in order to identify the crystalline phases involved in the composite fouling. Fig 5.2.11 shows the diffraction patterns of the composite scales together with that of the SiO_2 scale. The XRD spectra of all the scales gave a smooth broad peak or "halo" with a 2θ value of approximately 23°. This was caused by the presence of amorphous silica (Nakamura et al., 1989). However, this peak appeared to be narrower for the composite scales compared to that of the pure SiO_2 scale, probably due to the re-organization of SiO_2 molecules into a less disordered arrangement in the presence of COM, since the halo width represented the variation of the average distances of atoms in the amorphous Si-O-Si structural unit (Nakamura et al., 1989). The spectra of composite scales at 50 and 100 ppm COM were also found to contain additional sharp peaks (2θ values of 14.9^o, 24.4^o, 31.7^o, 38.3^o and 45.4^o). These are due to the presence of COM (Kontoyannis et al., 1997). The spectrum for 20 ppm COM composite scale did not show COM crystalline peaks, probably because of the low amount of COM formed (Table 5.2.2).

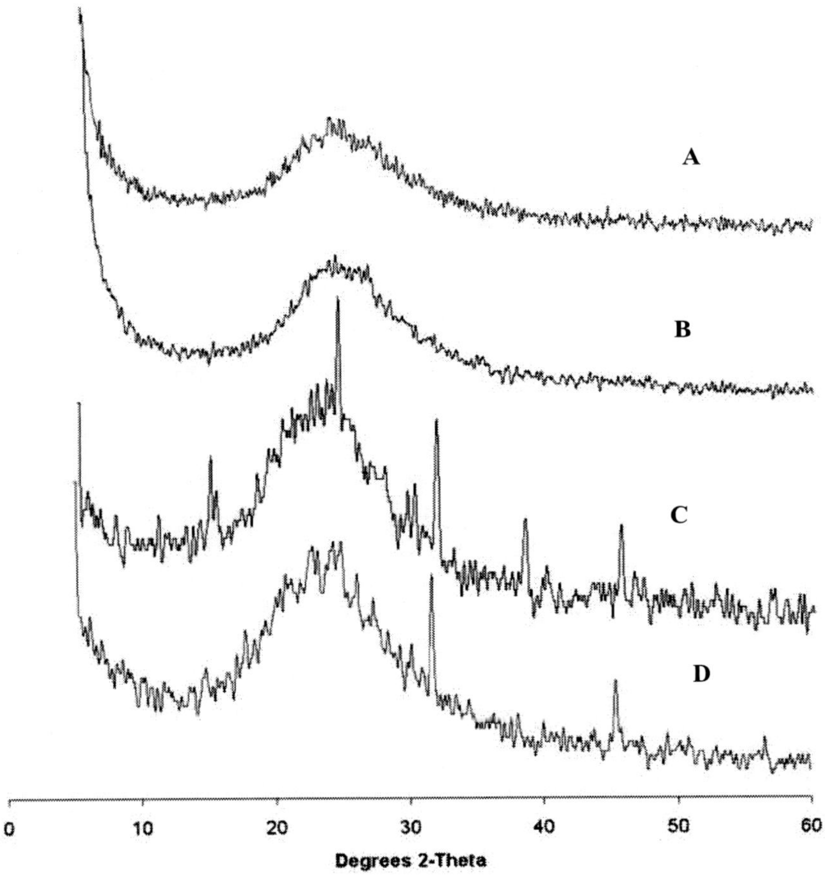

Fig. 5.2.11 X-ray diffraction patterns of (a) SiO_2 scale and composite scales, (b) initial COM 20 ppm, (c) initial COM 50 ppm, (d) initial COM 100 ppm

5.2.4 Effect of Boiling Heat Transfer

Under the forced convective and nucleate flow boiling heat transfer, bubble formation on certain nucleation sites of a heat exchange surface could be observed, especially at the upper end of the test tube, closest to the steam inlet. The increased heat transfer associated with bubble formation under boiling is primarily based on two mechanisms: microlayer evaporation and microconvection (Chang and Bankoff, 1990). Microlayer evaporation tends to promote scale deposition due to the local concentration effect as a thin liquid film, microlayer, beneath the bubble quickly evaporates. The detaching bubble then leaves behind a ring of deposit around the nucleation site from which the following bubbles can gradually build up a disc-shaped deposit (Freeborn and Lewis, 1962; Partridge and White, 1929b). Indeed this was observed in this study (Fig 5.2.12).

Table 5.2.2 Summary of experimental data and * molar ratios of SiO_2 and COM for the composite scales obtained from different initial COM concentrations

Initial COM/SiO_2 conc. (ppm)	Final COM conc. (ppm) δ	Final reactive SiO_2/total SiO_2 conc. (ppm) δ	Initial molar ratio SiO_2:COM in solution	Molar ratio SiO_2:COM at particle surface	** Molar ratio SiO_2:COM at particle cross-section
0/500	0	723.1/1500	--	--	--
20/0	22.8	0	--	--	--
50/0	31.8	0	--	--	--
100/0	27.1	0	--	--	--
20/500	76.3	925.1/1687.5	61:1	95:1	5:1
50/500	77.5	698.2/782.1	24:1	50:1	4:1
100/500	83.9	810.4/1789.3	12:1	34:1	4:1

* Quantitative results of the overall deposit composition are not available
** Average of at least ten measurements
δ see Appendix V for more details

Microconvection, on the other hand, can increase scale removal as a result of the extra stirring of the liquid boundary layer by detaching bubbles (Basset et al., 2000). It appeared that microconvection was the prevailing mechanism in this study since deposit accumulation was generally found to be thicker close to the steam outlet where bubble formation and bubble induced convection were less intense.

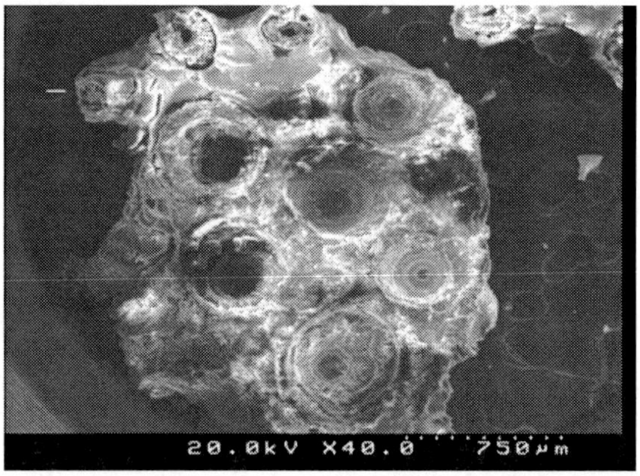

Fig 5.2.12 Scanning electron micrograph of bubble nucleation sites for composite scale (lower end)

5.2.5 Conclusions

The composite fouling of COM and SiO_2 has been studied under forced convective nucleate flow boiling heat transfer and evaporative conditions from both single and binary solutions with various COM/SiO_2 ratios. The results showed that the fouling resistance of pure COM began to rise immediately after the experiment started with the highest extent of fouling occurred at an initial COM concentration of 50 ppm (initial COM SS ~2.6). The fouling resistance of pure SiO_2 from an initial concentration of 500 ppm began to rise at a theoretical SS level of about 5.3. The overall fouling resistances of COM were found to be much less than that of SiO_2 due to rapid bulk precipitation. The process of SiO_2 deposition may involve different SiO_2 species at various degrees of evaporation and SiO_2 supersaturation, giving rise to the formation of two types of SiO_2 deposits with different physical characteristics.

The presence of COM at an initial concentration of 50 ppm led to a synergistic effect on composite fouling, which began at a lower theoretical SiO_2 SS level (~2.0) than that occurred for SiO_2 single system due to increased particulate deposition. In contrast, the composite fouling resistance in the presence of COM at an initial concentration of either 20 or 100 ppm was found to be lower than that of pure SiO_2. Part of this difference may be related to the changes in the magnitude of interfacial energy barrier between the surface of particle and wall and the physical properties of the fouling species such as particle size. The composite scale obtained from 50 ppm COM was found to be more tenacious than SiO_2 scale probably due to the cementing effect of COM on the deposit structure, as revealed by SEM analysis. The composite scales from 20 and 100 ppm COM were less tenacious than the 50 ppm COM composite scale due to the incomplete coverage on the tube surface. The boiling conditions might affect the scale deposition through microlayer evaporation and microconvection with microconvection being the prevailing factor.

5.3 COM-SiO_2 Composite Fouling in Sugar Solutions

In the previous section the composite fouling behaviour of COM and SiO_2 has been evaluated in aqueous solutions over a range of COM/SiO_2 supersaturation ratios to determine the critical feed composition at which the maximum degree of composite fouling occurs. This section describes the composite fouling characteristics of COM and SiO_2 from a synthetic juice solution under similar dynamic conditions. Results are presented for the fouling kinetics from mixed COM/SiO_2 sugar solutions at a range of initial supersaturation comparable to that used in aqueous solutions. The physico-chemical properties of scales obtained have been characterized using SEM-EDS, electrophoretic and particle size distribution measurements.

5.3.1 Composite Fouling Rate in Sugar Solutions

Experiments involving the composite fouling of COM/SiO_2 in the presence of sugar were evaluated over a comparable range of initial COM/SiO_2 SS ratios to that used in the absence of sugar so that direct comparison with previous data could be made (Fig 5.3.1 and Table 5.3.1).

As shown in Fig 5.3.1, fouling resistances obtained in the presence of sugar were similar to those achieved in water at lower CF ($\leq$~1.5), but began to rise sharply from a CF of ~1.6. This increase appeared reasonable as the presence of sugar could accelerate both silica polymerization and COM precipitation, the extent of which increased with increasing sugar concentrations (section 4.2). The deposition rates were also enhanced by higher SS levels of silica and COM in solution (Fig 5.3.1b/c) as a result of decreased solubilites of COM and silica with increased sugar concentration. The similar fouling rates observed at low CF levels, on the other hand, may be explained by an effect associated with hindered diffusion. An increase in the solution viscosity during progressive evaporation (Morison, 2002) may hinder particle transport and attachment to the fouling surface, which counter-balanced the impact of higher co-precipitation rates and SS of COM and SiO_2 at low CF but became less predominant at high CF levels. Fig 5.3.2 shows the effect of feed composition on the rate of composite fouling from sugar solutions. The composite fouling rates obtained at an intermediate COM SS (2.6) were significantly higher than that achieved at either low (1.1) or high (5.3) COM SS levels, coinciding well with a critical feed composition of maximum composite fouling rate identified in the previous study where no sugar was used (section 4.5).

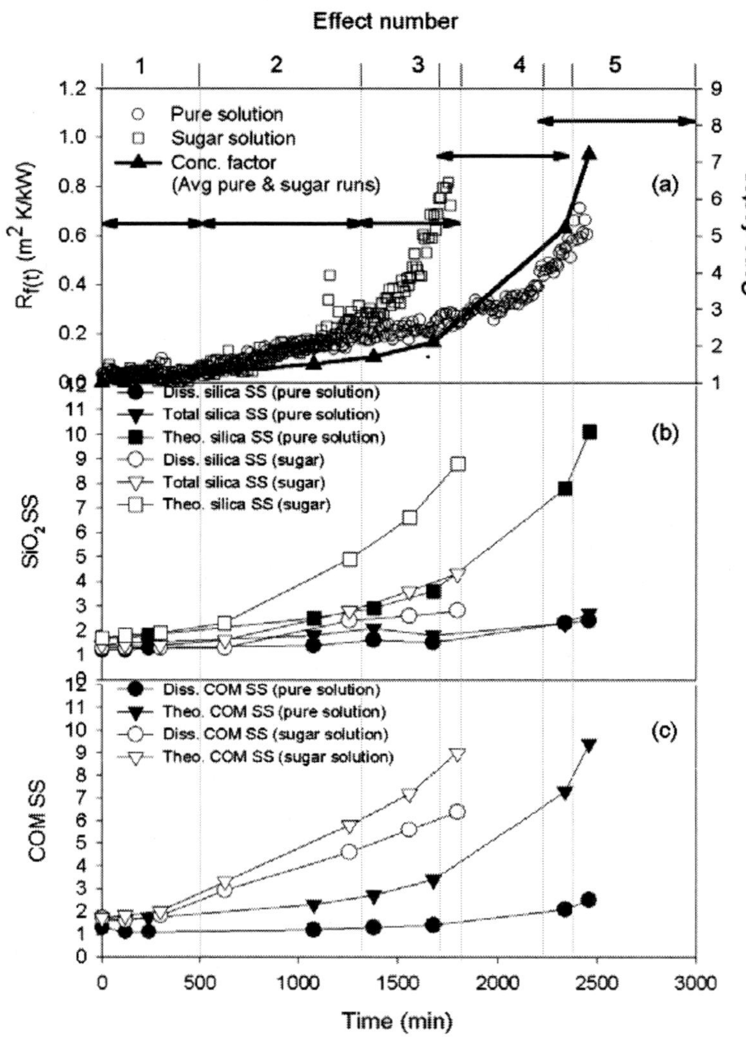

Fig. 5.3.1 Composite fouling of COM/SiO_2 in sugar solutions; initial sugar conc. 25%, initial COM SS 2.6, initial SiO_2 SS 1.7

Table 5.3.1 Summary of composite fouling data from sugar solutions (25%)

Initial COM/SiO_2 conc. (ppm)	Initial COM/SiO_2 SS in solution	Induction period (h)	CF at $R_{f(t)}$=0.2 (m^2 K/kW)
3.4/400	1.1/1.7	18.3	2.1
8.5/400	2.6/1.7	7.5	1.5
17/400	5.3/1.7	11.7	2.1

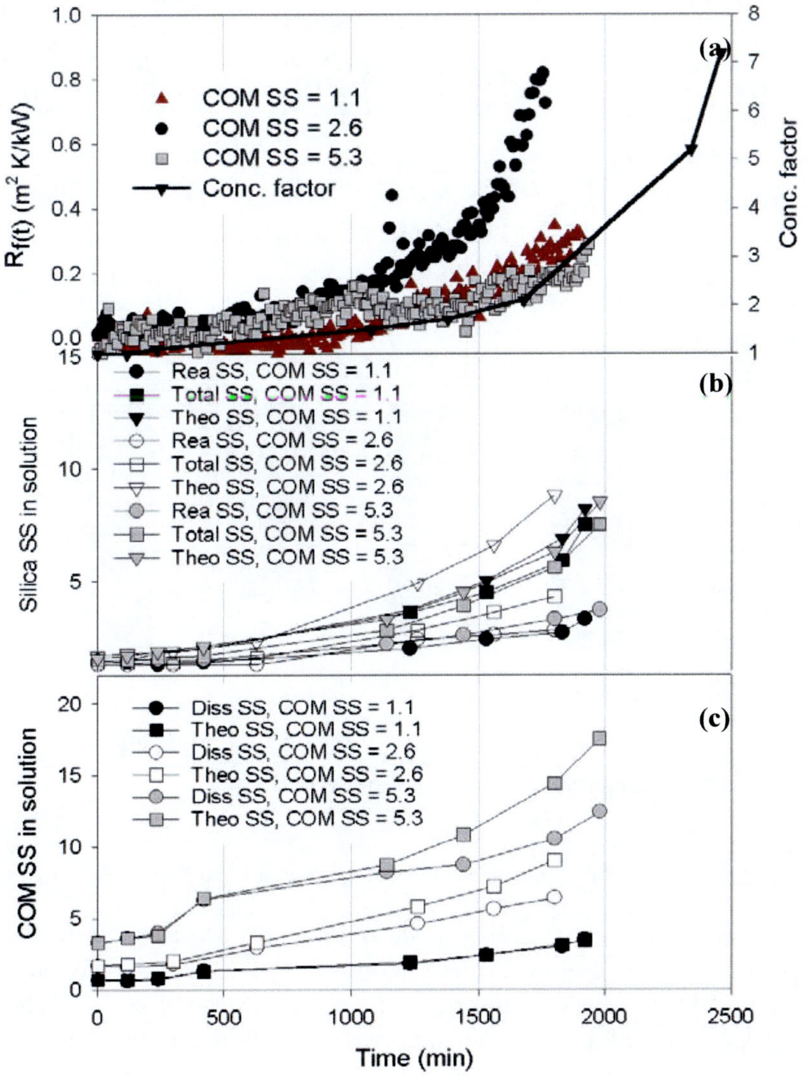

Fig. 5.3.2 Composite fouling of COM/SiO_2 in sugar solutions; initial sugar conc. 25%, initial SiO_2 SS 1.7

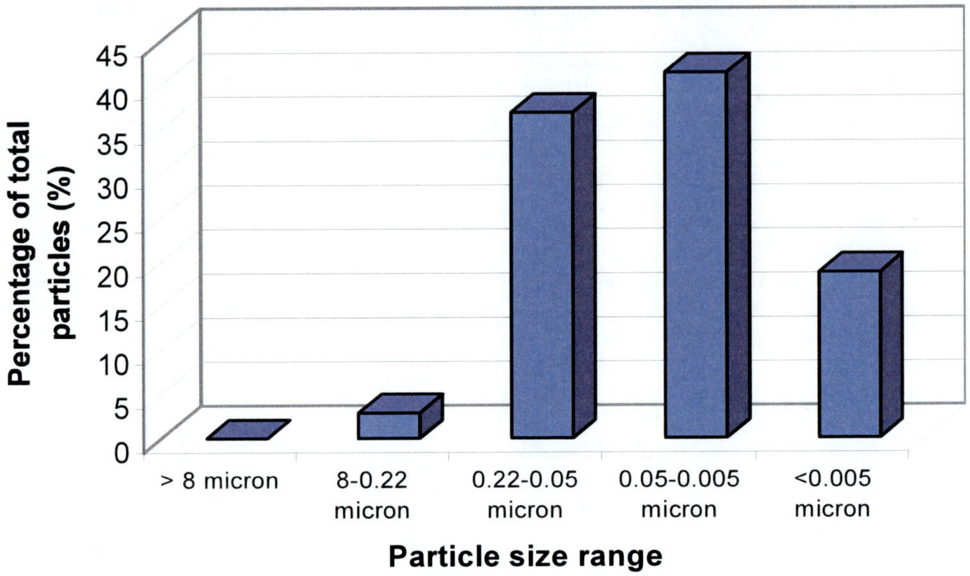

Fig. 5.3.3 Particle size distribution of COM-SiO_2 complexes formed in 25% sugar solution (measured by sequential filtration method from batch test)

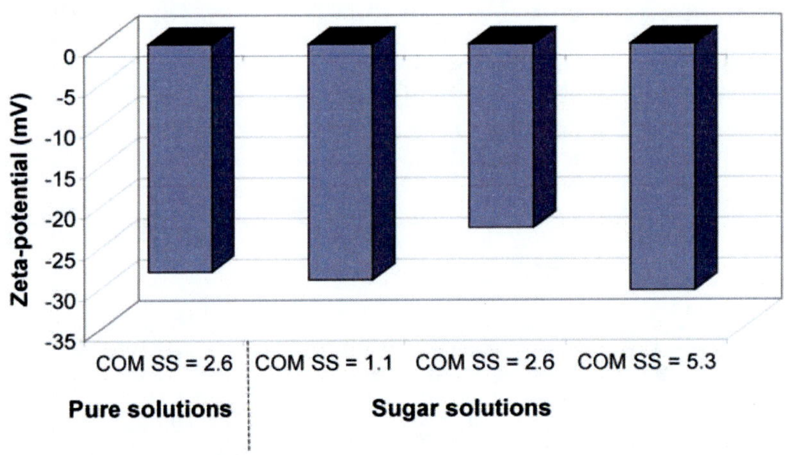

Fig. 5.3.4 Zeta-potential values of composite scales from pure and sugar solutions, initial SiO_2 SS = 1.7

The mechanism of composite fouling in sugar solution was further probed by measuring the zeta-potential values of resulting scales from both sugar and pure solutions and the size distribution of COM-SiO_2-sugar complexes formed in solution. Fig 5.3.3 shows that the species formed between COM, SiO_2 and sugar were mainly within the colloidal size range. The rate of composite fouling was shown to depend strongly on the zeta-potential values of sample particles, with the highest fouling rates corresponding to the lowest zeta-potential values (Fig 5.3.4). This was expected if the composite fouling was mainly due to the particulate deposition of colloidal particles in solution, as has been previously proposed (section 4.5): the composite scales were negatively charged at a solution pH of ~6.5. The surface of heat exchange tube (stainless steel) has an isoelectric point of between pH 3-4 and so was also negatively charged at this pH (Levy and Fritsch, 1959). Hence a reduction in the repulsive electrostatic double layer interactions between the tube surface and the colloidal particles containing COM, SiO_2 and sugar was expected to lower the interfacial energy barrier and favor their deposition onto the tube surface from solution. This hypothesis was supported by the fact that the amount of colloidal silica (i.e., the differences between reactive and total silica concentrations) remained in solution at an intermediate COM/SiO_2 SS ratio was smaller than that observed at either low or high COM/SiO_2 SS ratio (see Fig 5.3.2b).

5.3.2 Deposit Characteristics

Studies were conducted to characterize the physico-chemical properties of composite scales from sugar solutions using SEM-EDS and XRD analysis. The scanning electron micrographs (Figs 5.3.5a-c) show that the surfaces of all deposits appeared to contain conglomerates of micro-particles. Samples c and d (deposit obtained at high initial COM/SiO_2 SS ratio in sugar solution and from aqueous solution) also exhibited surface fragmentation and a layered structure possibly due to an increase in the crystallinity of the deposit.

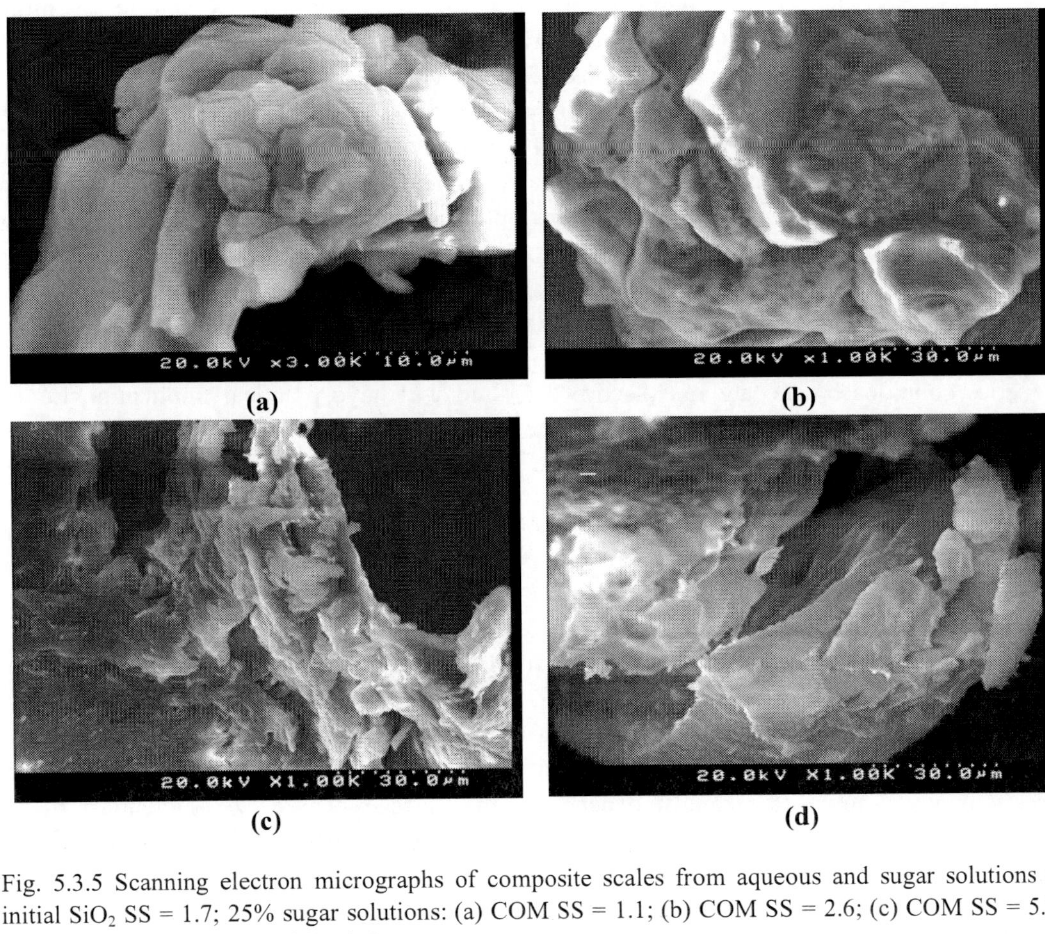

(a) (b) (c) (d)

Fig. 5.3.5 Scanning electron micrographs of composite scales from aqueous and sugar solutions at initial SiO_2 SS = 1.7; 25% sugar solutions: (a) COM SS = 1.1; (b) COM SS = 2.6; (c) COM SS = 5.3; aqueous solution: (d) COM SS = 2.6

EDS elemental analysis indicated that the composite scales were mainly composed of sugar (C and O), as anticipated based on its high initial concentration in solution (see Table

5.3.2), with minor amounts of Ca, Si and Cl. Concentration profiles determined by EDS quantitative analysis (Table 5.3.2) show that Ca and Si were mainly distributed in the interior of sample particles with an average Si /Ca molar ratio of ~2.0.

Table 5.3.2 EDS analysis of sucrose:Si:Ca molar ratios for the composite scales obtained from sugar solutions.

Initial COM/SiO_2 conc. (ppm)	Initial sucrose:Si:Ca molar ratio in solution	sucrose:Si:Ca molar ratio at particle surface	* sucrose:Si:Ca molar ratio at particle cross-section
3.4/400	33228:305:1	20.4:2.2:1	9.2:2.5:1
8.5/400	13291:122:1	24:2.1:1	9:2.3:1
17/400	6646:61:1	18.2:2.1:1	7.0:2.4:1

* Average of at least ten measurements

This value was lower than those obtained for scales formed in pure solutions (Table 5.2.2), possibly due to the dominant presence of sucrose molecules and lower initial concentrations of SiO_2 and COM in comparison to those in water. It has been suggested (Pannetier et al., 2001) that sucrose molecules form complexes of a coordination type with Ca, which generally has a total coordination number of 8. The molar ratios of sucrose:Ca at the interior of particles were found to be ~8-10, matching the literature data well. These data implied the possible formation of polymeric complex between COM, SiO_2 and sucrose followed by adsorption of sucrose molecules onto the complexes resulting in higher molar ratios of sucrose at particle surfaces.

Thus SEM-EDS analysis supported the evidence that particulate deposition involving complexes of COM, SiO_2 and sugar was the major mechanism during the composite fouling process of COM and SiO_2 in sugar solution. This confirmed previous results obtained from aqueous solutions by the authors (section 4.5) and will have practical implication for the development of pretreatment and control strategies as well as a physical model for composite fouling of COM and SiO_2. Further work to study the effects of operating conditions on composite fouling behaviour of COM and SiO_2 in sugar solutions is described in section 5.4.

5.3.3 Conclusions

This study has demonstrated that composite fouling of COM and SiO_2 was substantially enhanced by the presence of sugar in solution. The composite fouling rate at an intermediate COM SS (2.6) was found to be higher than those at either low or high COM SS, as was observed in aqueous solutions. The observed variability of composite fouling rates in sugar solutions was shown to strongly depend on the surface charge (zeta-potential) of the deposited scale particles, with the highest fouling rate corresponding to the lowest zeta-potential value of scale. This can be explained by a lower interfacial energy barrier between the surface of the heat exchanger tube and colloidal scale particles, both of which were negatively charged under the current experimental conditions. The mechanism of composite fouling of COM and SiO_2 was therefore found to be mainly associated with particulate deposition of polymeric COM-SiO_2-sugar complexes. The morphology of the composite scale may be altered due to the incorporation of COM.

5.4 Effect of Thermal Hydraulic Conditions on COM-SiO_2 Composite Fouling

It has been demonstrated in the previous sections (5.2 and 5.3) that the rate of composite fouling was dependent on solution composition, i.e., it was increased at an intermediate COM supersaturation (SS) level of 2.6 but was reduced at either low or high COM SS (1.1 and 5.3). The composite scale obtained at COM SS of 2.6 was also more tenacious than SiO_2 scale and the composite scales formed under other conditions tested. The primary mechanism of composite fouling was found to be particulate deposition of colloidal COM-SiO_2 complexes by overcoming the interfacial energy barrier between the heat exchange tube surface and that of COM-SiO_2 particle, both of which were negatively charged. The extent of fouling was surface-charge dependent with maximum fouling attained as the surface charges of the heat exchange tube and/or colloidal particle were minimized.

In order to successfully develop a predictive model simulating the composite fouling of COM and SiO_2, and to provide a guideline for optimal design, operation and control of sugar mill evaporators to minimize composite fouling, the effect of operating parameters on the fouling process needs to be known. The purpose of this section is to investigate the effects of thermal hydraulic operating conditions such as surface superheat, fluid velocity and bulk subcooling on the composite fouling process of COM-SiO_2 and deposit characteristics under a critical feed composition where maximum fouling occurred (see section 3.2.2.4). The physico-chemical properties of composite scales are also evaluated using instrumental analysis.

5.4.1 Effect of Superheat/Heat Flux on Fouling Rates

Under the subcooled flow boiling conditions employed in this work, the fouling rate, ϕ_f, was found to increase with heat flux, especially at high superheat levels (>10K) (Fig 5.4.1 and Table 5.4.1). The dependence of ϕ_f on superheat was approximately to the power of 3.3 [determined by fitting a power equation to data (Table 5.4.1)], which was comparable to those found for the deposition of hematite particles in boiling water systems (Charlesworth, 1970).

The increased composite fouling rate at higher heat flux may be related to enhanced bubble activities during nucleate boiling. Bubble-particle interactions have been investigated for oxides and silica by Leenaars (1988), mainly in order to characterize an approach of removing sub-micron particles by passage of an air/liquid mixture. The author observed that colloidal SiO_2 particles could be preferentially accumulated at the gas-liquid interface, provided that the wetting characteristics of particles were brought into proper range. Thus it was possible that, under certain conditions, the bubble nucleation sites act as particle binder promoting deposition. Furthermore, the evaporation of micro-layer water at vapor-liquid-solid triple interface of bubble nucleation sites (Fig 5.4.2), which was proposed for the deposition of silica and corrosion products in boiling water systems (Nishino et al., 1991; Pulling and

Collier, 1963; Tomlinson et al., 1990), may also be a significant factor here. Micro-layer evaporation might not only cause a local increase in the concentration of COM and SiO_2 at bubble nucleation sites (Burrill, 1979; Tomlinson et al., 1990), but also enhance the strength of composite scales, which, as shown later, was the result of deposit cementation by COM at heating surface.

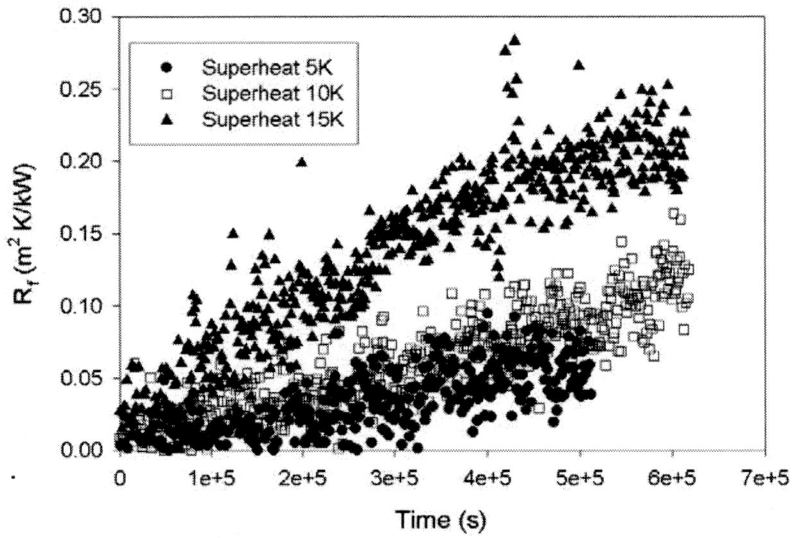

Fig. 5.4.1 Effect of superheat on the rate of composite fouling in sugar solution, velocity 1.1 m/s, subcooling 20 K, sugar conc. 25%, conc. factor ~1.7

Table 5.4.1 Summary of composite fouling rate data for various thermal hydraulic conditions, sugar conc. 25%, COM SS ~4.4, SiO_2 SS ~2.9. conc. factor ~1.7 (Appendix VI)

Run No.	Superheat (K)	Heat flux (kW/m^2)	Velocity (m/s)	Subcooling (K)	* ϕ_f (x10^{-7} (m^2 K/kW·s))
1	5	56.4	1.1	20	1.2
2	10	67.9	1.1	20	1.8
3	15	80.4	1.1	20	4.1
4	10	48.0	0.8	20	4.2
5	10	79.4	1.4	20	1.7
$^{\alpha}$6	10	79.4	1.1	25	3.2
$^{\alpha}$7	10	87.7	1.1	30	2.4
8	10	104.9	1.8	20	1.7

* fouling rate determined by linear regression

$^{\alpha}$ at comparable initial COM/SiO_2 supersaturations to those of the other runs because of different solubilities of COM and SiO_2 in solution resulting from changes in the bulk temperature.

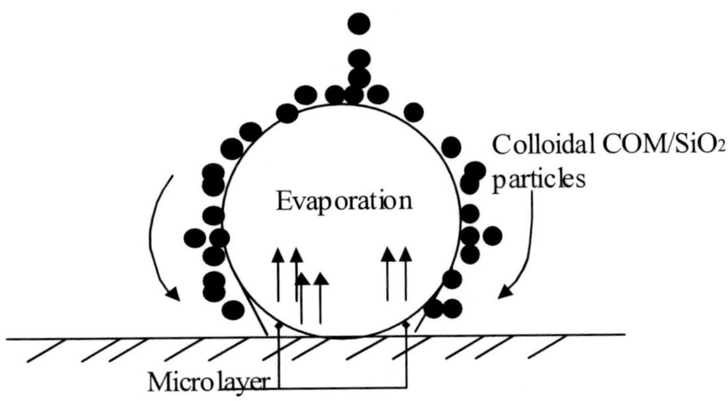

Fig. 5.4.2 Schematic deposition model of micro-layer evaporation at bubble nucleation site

The fouling curve generally exhibited linear increase (Fig 5.4.1) within the duration of the experiment, implying a constant rather than increased removal rate with the growing deposit. This behaviour was not surprising based on the tenacious nature of the composite deposit (section 4.5). Assuming an adhesion-controlled deposition process for the range of Reynolds number used in this work (see Table 3.3 and section 4.5) and 100% particle attachment efficiency, the rate of particle deposition could be expressed as:

$$\phi_d \approx \frac{K_a C_b}{\rho_f k_f} \qquad (5.4.1)$$

where K_a is the adhesion rate coefficient (m/s) and C_b is the bulk concentration of particles (kg/m^3). ρ_f and k_f are the density (kg/m^3) and thermal conductivity (W/m K) of deposit, respectively (Epstein, 1999; Oliveira et al., 1993). The adhesion rate coefficient is related to surface temperature, T_s (K), by an Arrhenius type dependence analogous to a chemical reaction:

$$K_a = K_o \exp(-E / RT_s) \qquad (5.4.2)$$

where K_o is the pre-exponential factor and E is the activation energy (kJ/g mol), which is a measure of bonding strength of deposit (Turner, 1993). The activation energy was estimated to be ~128 ± 15 kJ/g mol (see Appendix VIB). This was of similar magnitude to typical chemical bond strength, and was considerably higher than the literature values typical of silica scales (~40-85 kJ/g mol, surface temperature-based) (Jamialahmadi and Muller-Steinhagen, 1993). Hence the higher bonding strength of the composite deposit was in good agreement with previous experimental evidence that CaOx-SiO_2 composite scale was more tenacious than pure silica scale (section 4.5) due to the cementing effect of CaOx.

It was proposed by Nicholson and Sarbutt (1980), Hussain et al. (1986), and Turner and Klimas (2001) that deposit consolidation experienced by scales formed under high temperatures on the boiling surface might result in an increase in the bond strength of the deposit, making it hard to remove with low shear forces. Consolidation is a process in which particles are attached to the heat transfer surface or pre-existing deposit by chemical bonding.

The driving force of consolidation varies according to the type of fouling species present and may include Ostwald ripening and boiling induced precipitation of dissolved species (Turner and Klimas, 2001). For COM-SiO_2 composite fouling, consolidation might have been achieved by the cementing effect of calcium oxalate precipitation in the silica matrix through micro-layer evaporation (see Fig 5.4.2). Consequently, the rate and extent of consolidation will depend on the rate of evaporation or heat flux.

5.4.2 Effect of Subcooling on Fouling Rates

Fig 5.4.3 shows the effect of bulk subcooling on composite fouling under a constant superheat (10 K) and fluid velocity (1.1 m/s). The rate of composite fouling was markedly higher at a bulk subcooling of ~25 K in comparison to those obtained at 20 and 30 K (Table 5.4.1).

The initial increase in fouling rate was consistent with previous studies of particle deposition in subcooled boiling water (Lister, 1981), which may be due to enhanced deposit consolidation at high heat flux (Table 5.4.1).

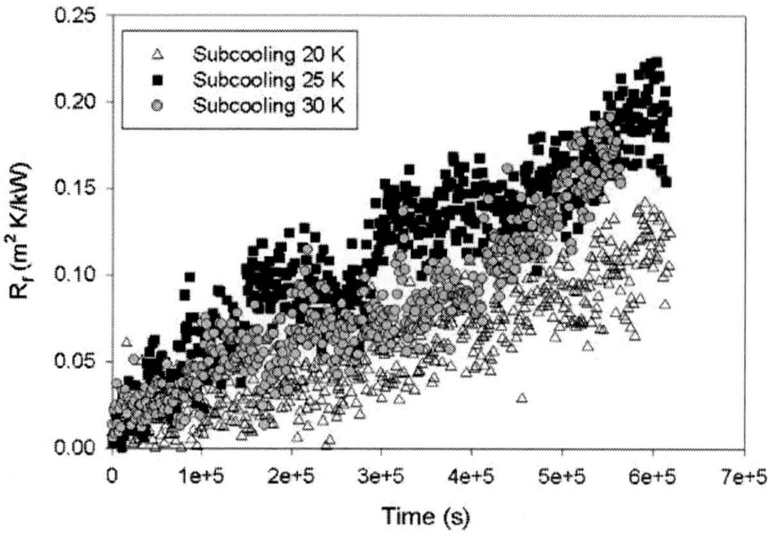

Fig. 5.4.3 Effect of bulk subcooling on the rate of composite fouling in sugar solution, surface superheat 10 K, fluid velocity 1.1 m/s, sugar conc. 25%

An increase in the bulk subcooling, however, would also lead to changes in the physical properties of liquid suspension such as viscosity. It was established that near-wall lift force or Saffman lift force (Epstein, 1997; Lister, 1981), a component of the hydrodynamic forces governing particle re-entrainment, had a linear dependency on the fluid viscosity (Williams et al., 1996). So the drops in the fouling rate at higher subcooling may be the result, at least in part, of an increased hydrodynamic lift forces, which prevented adhesion and counter-balanced the effects of deposit consolidation under increasing heat flux.

5.4.3 Effect of Velocity on Fouling Rates

The effect of fluid velocity on the rate of composite fouling ϕ_f at a superheat of 10 K and a bulk subcooling of 20 K is shown in Figs 5.4.4-5.4.5. An increase in the fluid velocity reduced the rate of fouling, although only a slight decrease in ϕ_f could be observed for velocities above 1.1 (m/s).

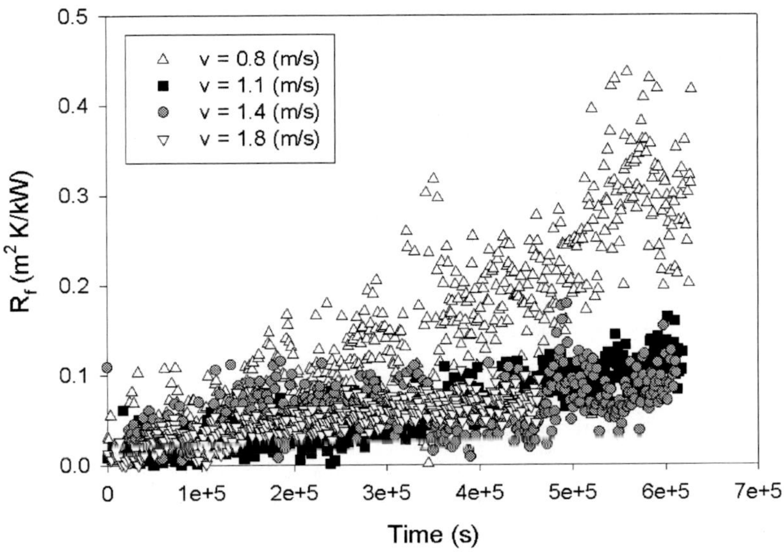

Fig. 5.4.4 Effect of fluid velocity on the rate of composite fouling in sugar solution, surface superheat 10 K, bulk subcooling 20 K, sugar conc. 25%.

The inverse dependence of fouling rate on velocity correlated well with the general assumption that COM/SiO_2 composite fouling was a surface adhesion-controlled process (section 4.5) – the deposition rate was independent of fluid velocity (eqn. 5.4.1) while the removal rate increased with increasing shear and drag forces at the wall (Epstein, 1997; Oliveira et al., 1993). Taking the influences of hydrodynamic lift forces into account, recent studies of particle deposition from flowing water suspension have actually shown that the attachment efficiency to the surface also tended to decrease with the wall shear stress (Adomeit and Renz, 2000; Yiantsios and Karabelas, 2003). Hence the effect of fluid velocity may involve a dual role of enhancing removal forces as well as inhibiting the attachment of particle.

The little variation in the fouling rate for velocities 1.1 – 1.8 (m/s) (Fig 5.4.5) may be attributable to an increase in the rate of consolidation and hence deposit strength with shear stress (Melo and Pinheiro, 1986; Oliveira et al., 1988; Turner and Klimas, 2001), which probably compensated the tendency of high velocity to reduce the net rate of fouling. Melo and Pinheiro (1986) investigated the particulate fouling of kaolin-water systems in an annular heat exchanger. They showed that the kaolin deposit consisted of a two-layer structure with the inner layer being more cohesive than the outer layer. The deposit strength was found to increase with increasing Reynolds number because of a rising proportion of hard layer formed in the deposit. This phenomenon was supported in the present study by results of the instrumental analysis on the deposits as well as visual and tactile study showing a bilayer structure in the deposit – a hard inner layer and a soft outer layer, as will be discussed in the next section.

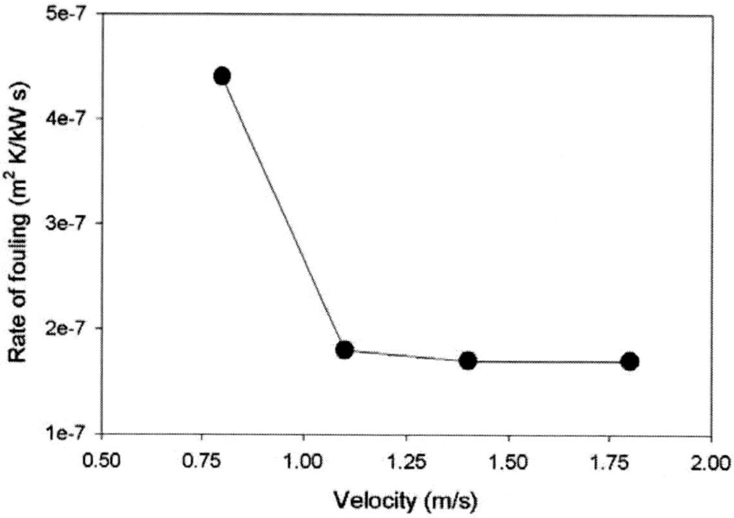

Fig. 5.4.5 Changes in the fouling rate as a function of fluid velocity, surface superheat 10 K, bulk subcooling 20 K, sugar conc. 25%.

5.4.4 Deposit Characteristics

Visual and tactile inspection of the deposition process during and after the fouling experiments suggested the existence of two layers of different characteristics – a hard inner layer (~0.10 mm thickness) covered by a soft outer layer (~0.15 mm thickness), the latter being readily dislodgeable by the shear force of flowing liquid during the run. This deposit bilayer was later confirmed by SEM analysis (Fig 5.4.6), which showed that the outer layer had a porous structure containing aggregates of large particles, while the inner layer appeared to be much more compact and homogenous.

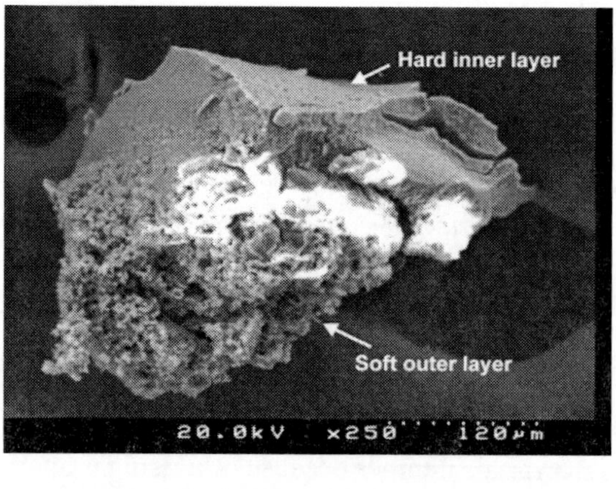

Fig. 5.4.6 The two-layer structure of composite deposit (Run 6)

Generally, deposits obtained at low surface superheat (5-10K), velocity (0.8-1.1m/s) and bulk subcooling (20K) (Figs 5.4.7A-C) appeared to be more porous than those formed in the presence of high superheat (15K), velocity (1.4 -1.8 m/s) or subcooling (25-30K). The latter exhibited higher tenacity (upon removal from the tube surface) and disc- or ring-shaped morphologies (Figs 5.4.7D and F) with primarily cohesive structures and traces of loose components on the edges and surfaces (Figs 5.4.7D-H). The observed structural transformation supported the general argument of deposit consolidation by COM via micro-layer evaporation under high heat flux, which increased the fraction of the compact layer formed in the deposit. Furthermore, the results of the XRD analysis indicated the presence of both COM and COD together with SiO_2 as the main ingredient, consistent with data obtained previously from batch co-precipitation tests (see section 4.3).

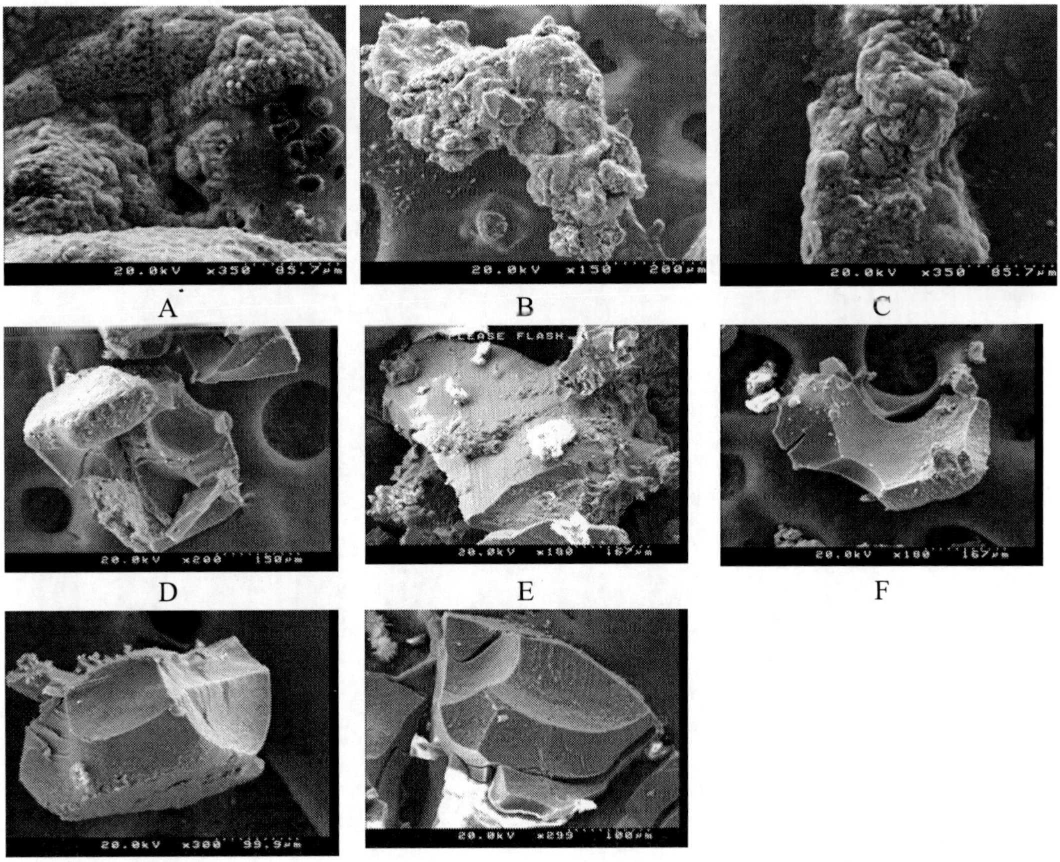

Fig. 5.4.7 SEM micrographs of the composite deposits obtained under different thermal hydraulic conditions, sugar conc. 25%; (A) Run 1, (B) Run 2, (C) Run 4, (D) Run 3, (E) Run 5, (F) Run 6, (G) Run 7, (H) Run 8

EDS elemental analysis was used to determine the chemical composition of the deposit as a function of thermal hydraulic conditions under which the composite deposits formed, as shown in Fig 5.4.8. The data are presented as Si:Ca molar ratios at the cross-sections of sample particles, which reflected the extent of deposit consolidation by COM and hence the deposit strength. It is evident that the deposit strength increased with increasing surface superheat, velocity and bulk subcooling, which matched the trend of structual transformation from SEM observations (Fig 5.4.7). As indicated earlier, the influences of these operating parameters on deposit structure were *in essence* a consequence of corresponding changes in the heat flux of the system, and potential models that could appropriately describe the deposit consolidation data needed at least to include such an effect. For this reason, a consolidation factor, β_{CaOx}, has been defined in order to quantify the effects of heat flux on the extent of deposit consolidation:

$$\beta_{CaOx} = \frac{1}{M_R} \tag{5.4.3}$$

where M_R is the Si/Ca molar ratio of deposit cross-sections obtained under different superheat values. Fig 5.4.9 shows a linear relationship between deposit strength and heat flux.

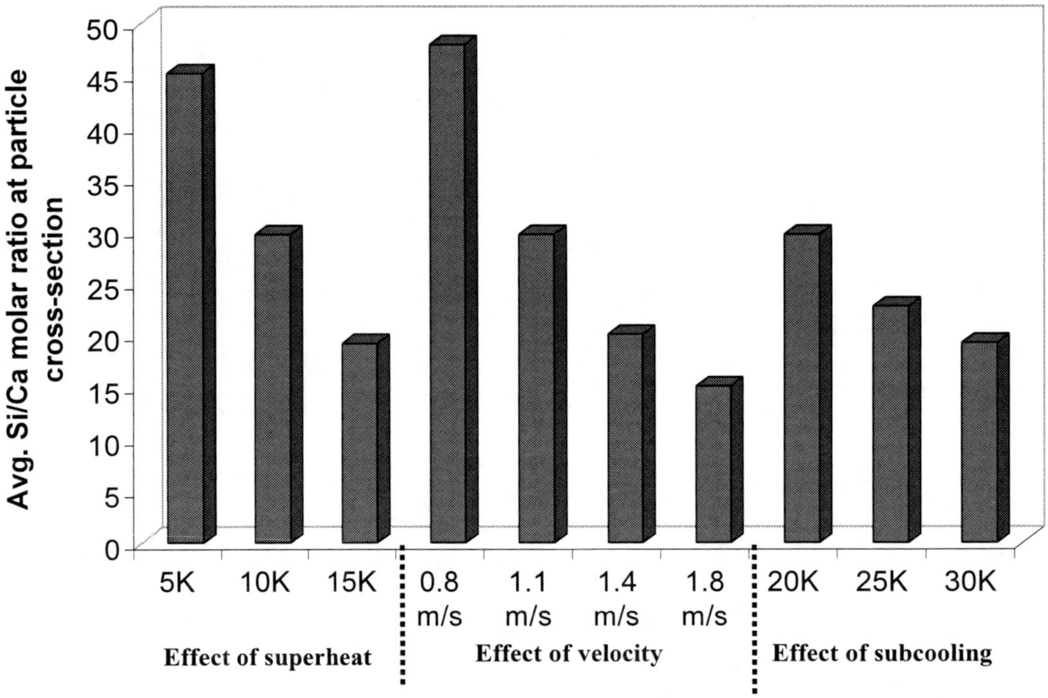

Fig. 5.4.8 EDS analysis of Si/Ca molar ratios at particle cross-sections of composite scales from various operating conditions

It was of interest to note that at a superheat of 10K, an increase in the velocity from 1.1 to 1.8 m/s was seen to increase the deposit strength, β_{CaOx} (Fig 5.4.8), while the fouling rate remained relatively constant (Fig 5.4.5 and Table 5.4.1). These results implied that velocity adjustment towards higher range (> 1.1 m/s) during sugar processing would incur additional costs of energy consumption and cleaning due to the higher tenacity of scales without having much impact on the rate of composite fouling in the evaporator. It should be recognized, however, that the actual structural transformations of the composite deposits could be much more complicated in sugar factories; if the presence of other non-sugar organic species and changes in the sugar concentration significantly altered the composite fouling behavior and deposit strength, for instance, then the measured threshold velocity of 1.1 m/s obtained in this work would simply have little practical value.

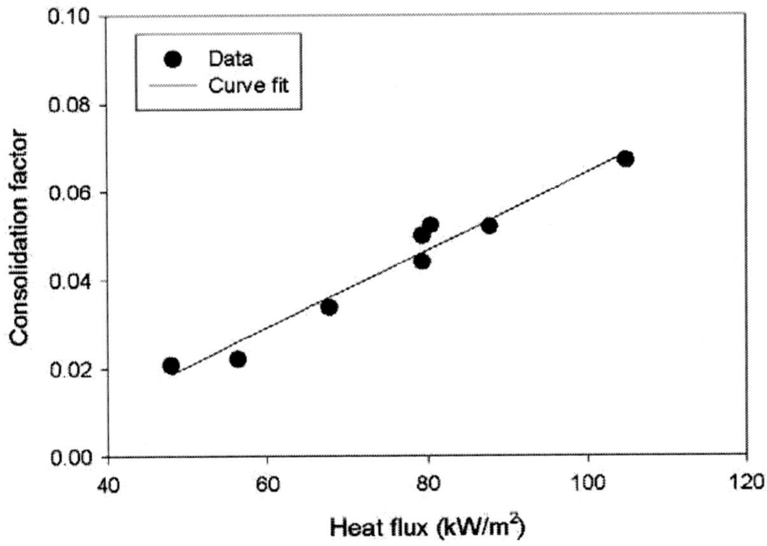

Fig. 5.4.9 Effect of heat flux on deposit consolidation

5.4.5 Conclusions

The results presented in this section have demonstrated the effects of operating conditions on COM/SiO_2 composite fouling and the nature of the scales formed in synthetic juice solutions. Three key operating parameters, surface superheat, fluid velocity and bulk subcooling, were studied under a constant solution composition. The linear fouling rate was strongly dependent on the surface superheat, with the fastest fouling observed for the highest superheat. For a constant superheat of 10 K, the linear fouling rate was found to exhibit a maximum at a subcooling of ~25 K, but remained constant for fluid velocities 1.1 m/s or higher. These changes may be explained by bubble-particle interactions at the boiling metal surface, hydrodynamic lift forces at the wall the fouled tube and structural consolidation of the composite deposit. The composite scales formed under the current thermal-hydraulic conditions were shown to have a high bonding strength (~128 kJ/g mole), possibly due to the structural consolidation by CaOx via microlayer evaporation. The extent of consolidation was found to increase with surface superheat, velocity and bulk subcooling.

Chapter 6

MODELING OF COMPOSITE FOULING OF CAOX-SIO_2

For evaluating and obtaining strategies to effectively mitigate the composite fouling of COM and SiO_2, it is essential to be able to make predications of the rate of composite fouling. Good predictions will also facilitate the development of improved design and operation of the evaporator station. In order to meet this practical need, this section presents an approach to modeling the COM-SiO_2 composite fouling based on the effects of various thermal hydraulic operating parameters, as well as some of the mechanistic assumptions previously proposed (see section 4.7). In particular, deposit consolidation by COM in the matrix of SiO_2 scales is considered through microlayer evaporation. The model predictions are compared with experimental data for a constant solution composition and pH (i.e., initial COM and SiO_2 supersaturations and sugar concentration). Factors that influence key concepts and parameters of the composite fouling model are briefly discussed.

6.1 PREVIOUS MODELING STUDIES ON PARTICULATE AND SILICA FOULING

Most mathematical models of particulate fouling applied in the literature are based on a general equation proposed by Kern and Seaton (1959):

$$\frac{dR_f}{dt} = \phi_d - \phi_r \tag{6.1}$$

where R_f is the thermal or fouling resistance of the deposit, ϕ_d is the deposition rate and ϕ_r is the removal rate. Considering a constant deposition rate and an increasing removal rate with the mass of deposit, the integrated form of eqn (6.1) may be expressed by:

$$R_f = R_f^*(1 - e^{-Bt}) \tag{6.2}$$

where R_f^* (= ϕ_d/B) is the asymptotic fouling resistance and B is a constant related to deposit strength. The deposition rate was proposed to be a function of bulk concentration of fouling species, C_b, and fluid velocity, u, while the removal rate was dependant on wall shear stress, τ, and deposit thickness, x_f:

$$\phi_d = K_1 C_b u \quad (6.3)$$

$$\phi_r = K_2 \tau x_f = K_3 u^2 R_f \quad (6.4)$$

where K_1, K_2 and K_3 are constants. The basic concept of Kern – Seaton (eqn 6.1) has been widely used for the development of fouling models such as the Watkinson and Epstein model, the Gudmundsson model, the Epstein model and the Melo & Pinheiro model, which have different functional expressions of ϕ_d and ϕ_r (eqns 6.5 – 6.12, Table 6.1) depending on the specific fouling species and mechanisms involved. Features of these models have been reviewed by Epstein (1997, 1999) and Pinheiro (1986).

Table 6.1 Summary of fouling models derived from Kern – Seaton concept

Model	Deposition term	Removal term	Applicable conditions
Watkinson & Epstein model (1970)	$\phi_d = NP = K_4(C_b - C_s)P$ (6.5) $P = \frac{K_5 e^{\left(-\frac{E}{RT_s}\right)}}{u^2 f}$ (6.6)	same as the Kern – Seaton model (eqn 6.4)	Particulate fouling
Gudmudsson model (1979)	same as above	$\phi_r = K_6 u R_f$ (6.7)	Particulate fouling, also correlates gas-oil fouling data
Taborek et al. model (1972a)	$\phi_d = K_7 P_d \Omega^i e^{\left(-\frac{E}{RT_s}\right)}$ (6.8)	$\phi_r = K_8 \frac{\tau}{\psi} x_f$ (6.9)	$CaCO_3$ crystallization fouling in the presence of particulate matters
Melo & Pinheiro model (1986)	$\phi_d = \frac{C}{(1/K_t + 1/K_r)\rho_f k_f}$ (6.10)	$\phi_r = K_9 u^{2-\varsigma} R_f$ (6.11)*	Generalized model
Epstein model (1997)	$\phi_d = \frac{C}{\frac{k' Sc^{2/3}}{u^*} + \left[\frac{k'' e^{E/RT_s}(u^*)^2}{\nu \phi_d^{i-1}}\right]^{1/i}}$ (6.12)	n/a	Initial chemical reaction and particulate fouling, verified for colloidal particles and surface charge effects

* $\zeta \approx 1$ for kaolin-water systems at high Reynolds numbers (7000-12000), where surface adhesion controlled the deposition process.

Turner and Klimas (2001) proposed a particulate fouling model to include the effect of deposit consolidation on the kinetics of fouling under flow boiling conditions.

$$\frac{dm_t}{dt} = K_d C - K_{rem} m_r(t) \tag{6.13}$$

$$\frac{dm_r}{dt} = \frac{dm_t}{dt} - K_{con} m_r(t) \tag{6.14}$$

where m_t is the total deposit mass, m_r is the part of the total deposit mass available to be removed and C is the concentration of particles in solution. K_d, K_{rem} and K_{con} are the deposition, removal and consolidation rate coefficient, repectively. Eqns (6.13) and (6.14) could be combined and integrated to the following expression of m_t:

$$m_t = \frac{K_d C}{K_{rem} + K_{con}} \left(tK_{con} + \frac{K_{rem}}{(K_{rem} + K_{con})^2} (1 - e^{[-(K_{rem}+K_{con})t]}) \right) \tag{6.15}$$

Thus the model assumed the existence of a deposit bilayer consisting of a consolidated inner layer surrounded by an unconsolidated loose layer, and was capable of predicting different fouling kinetic profiles depending on the relative magnitudes of K_d, K_{rem} and K_{con}. For high rates of deposit consolidation, the above equation simplified into a linear kinetics form:

$$m_t = K_d C \frac{K_{con}}{K_{rem} + K_{con}} t \tag{6.16}$$

The authors proposed that consolidation be considered a process in which chemical bond was formed between the particle and the heat transfer surface or pre-deposited film. Potential mechanisms for deposit consolidation included Ostwald ripening, dissolution – recrystallization and boiling-induced precipitation of dissolved species in the pores of the deposit. Therefore the rate of consolidation, especially under boiling conditions, was strongly affected by factors such as heat flux.

The effect of heat flux on the deposition of iron oxide particles on boiling surface has been demonstrated by Asakura et al. (1978), who formulated a model based on micro-layer evaporation:

$$\frac{dm_t}{dt} = \frac{Kq''C}{h_{fg}} \tag{6.17}$$

where q" is the heat flux, h_{fg} is the latent heat, C is the particle concentration in solution and qC/h_{fg} is the amount of particles contained in the vaporized water during nucleate boiling. K is the deposition rate coefficient representing the fraction of particles that deposited on the surface.

$$K = \left(\frac{4}{3}\right)^2 R_b U_b \frac{\rho_G^2}{\rho_l} v_l \tag{6.18}$$

where R_b is the maximum bubble radius and U_b is the bubble growth rate.

Chan et al. (1988) derived an analytical model which could predict the deposition of silica from geothermal brines in tubular heat exchangers as a function of silica supersaturation and pH:

$$\frac{dR_f}{dt} = K_R \left(\frac{C_b - C_e}{C_e}\right)^p \left(C_{OH^-}\right)^r \tag{6.19}$$

where K_R is the deposition rate constant, C_b is the bulk concentration of silica, C_e is the silica solubility at surface temperature, C_{OH^-} is the molar concentration of OH^- ions in the bulk solution and p is the reaction order of silica. The thermal hydraulic conditions were accounted for in the model by evaluating their effects on surface temperature.

6.2 Composite Fouling Model

The model used for the present study was developed based on the concept of Turner and Klimas model (2001) (eqn 6.16) since it included the component of deposit consolidation. Consolidation has been found to be a significant mechanism affecting the rate of composite fouling and the structure of composite deposits. Thus the eqn 6.16 was modified as:

$$R_f = \frac{K_d C_{SiO2}}{\rho_f k_f} \frac{\phi_c}{\phi_r + \phi_c} t \tag{6.20}$$

where ρ_f and k_f are density and thermal conductivity of deposit, respectively. C_{SiO2} is the concentration of colloidal silica in solution, which is the major source of particles in solution. The determination of deposition, removal and consolidation terms (i.e., K_d, ϕ_r and ϕ_c) were described below.

(i) Deposition The Watkinson and Epstein model and the Pinheiro model were used to formulate the deposition term. Under the operating conditions tested in this work, deposition was a surface-controlled process. This could be expressed in a modified deposition term of Watkinson model for adhesion control by assuming that the interfacial energy barrier for particle attachment was applicable to boiling conditions:

$$K_d = \frac{A e^{\left(-\frac{E}{RT_s}\right)}}{u^2} \tag{6.21}$$

where A is a pre-exponential factor including the effects of all the other variables apart from T_s (Pinheiro, 1986). K_d was considered to have a strong velocity dependancy similar to those demonstrated for the deposition of kaolin and silica under turbulant flow conditions (Melo and Pinheiro, 1986; Yiantsios and Karabelas, 2003).

(ii) Removal The removal term was derived based on the Pineiro model as follows:

$$\phi_r = K' u^{2+b} \tag{6.22}$$

where b is a constant comprising the effect of enhanced hydrodynamic lift forces due to changes in fluid viscosity (subcooling effect) (section 4.7). K' is a coefficient with a similar physical meaning to A (i.e., comprising effects other than those of velocity and subcooling (Pinheiro, 1986)). If a linear relationship was assumed between the removal rate and fluid viscosity (Williams et al., 1996), then b=1 and the above equation becomes:

$$\phi_r = K' u^3 \tag{6.23}$$

(iii) Consolidation The Asakura et al. model was used to define the rate of consolidation during composite fouling by considering COM precipitation through micro-layer evaporation only under constant heat flux operation mode (section 4.7), viz,

$$\phi_c = \frac{C_{COM} q''}{h_{fg} + c_p \Delta T_{sub}} K_{con} \tag{6.24}$$

where q" is the heat flux, C_{COM} is the concentration of COM in solution, c_p is the fluid specific heat, ΔT_{sub} is the bulk subcooling and K_{con} has the same expression as eqn (6.18). The values of R_b and U in eqn (6.18) were determined from published data (Asakura et al., 1978). The composite fouling resistance in terms of thermal hydraulic operating conditions was obtained by combining eqns (6.20) – (6.24), thus,

$$R_f = \frac{A \exp(-E/RT_s) C_{SiO2}}{u^2 k_f \rho_f (K' M^* + 1)} t \tag{6.25}$$

where M* is given by eqn 6.26 as:

$$M^* = \frac{u^3 (h_{fg} + c_p \Delta T_{sub})}{K_{con} q'' C_{COM}} \tag{6.26}$$

With the above two equations and eqn (6.17) it is possible to determine the composite fouling rates as a function of heat flux, fluid velocity and bulk subcooling. Other necessary input parameters include concentrations of COM and SiO_2 and thermo-physical properties of liquid and steam (which can be determined directly based on surface or bulk temperature). Only A and K' have to be determined through individual data set fitting (i.e., eqn (6.25)).

6.3 Comparison of Models with Composite Fouling Data

The Chan et al. silica deposition model (eqn 6.19), the Watkinson particulate fouling model (eqns 6.4-6.6), and the model proposed in the present study (eqns 6.25 and 6.26) were compared with a limited amount of composite fouling data. The fitted values of A and K' for the composite fouling model are summarized in Table 6.2.

Table 6.2 Parameter values of composite fouling model for $CaOx$-SiO_2 binary systems in sugar solutions

Run No.	Superheat (K)	Heat flux (kW/m^2)	Velocity (m/s)	Subcooling (K)	A ($x10^{-5}$)	K' ($x10^{-17}$)	A/K' ($x10^{11}$)
1	5	56.4	1.1	20	3.4	9.9	3.4
2	10	67.9	1.1	20	5.9	13.0	4.5
3	15	80.4	1.1	20	12.8	14.3	9.0
4	10	48.0	0.8	20	3.3	200.0	1.7
5	10	79.4	1.4	20	2.2	3.6	6.1
6	10	79.4	1.1	25	10.0	14.0	7.1
7	10	87.7	1.1	30	120	200.0	6.0
8	10	104.9	1.8	20	1.9	1.7	11.2

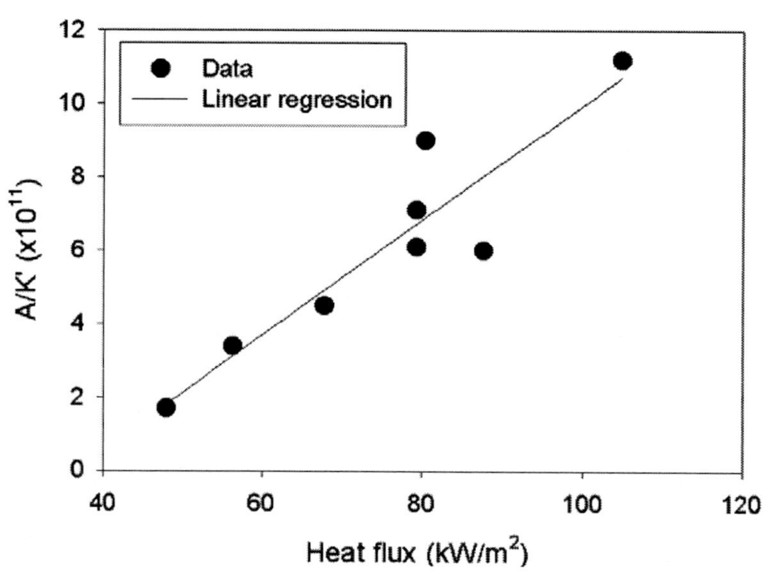

Fig. 6.1 Changes in the calculated parameters for the composite fouling model as a function of heat flux

It is apparent that the ratio between A and K' linearly increased from $1.7x10^{11}$ at heat flux 48 kW/m^2 to $11.2x10^{11}$ at heat flux 105 kW/m^2 (Fig 6.1), reflecting a higher degree of deposit consolidation at higher heat flux level. Somewhat similar trend was observed (Fig 6.2) for K_R values in the Chan et al. model (eqn (6.19)) except that the latter gave more scatter around linearity, presumably because the effects of thermal hydraulic parameters other than heat flux were not considered in the Chan et al. model. The value of r in the Chan et al.

model was found to vary in the range of 0.8-0.95, which was in reasonable agreement with the published data of 1.0 (Chan et al., 1988). The parameter values of K_d/K_3 in the Watkinson and Epstein model (eqn (6.4) and (6.5)) are shown in Fig 6.3. In contrast to the trend given by the composite fouling model, the linear regression indicated reduced K_d/K_3 at high heat flux levels. The accuracy of these results was compromised by the restricted number of data points produced from closed-loop fouling runs.

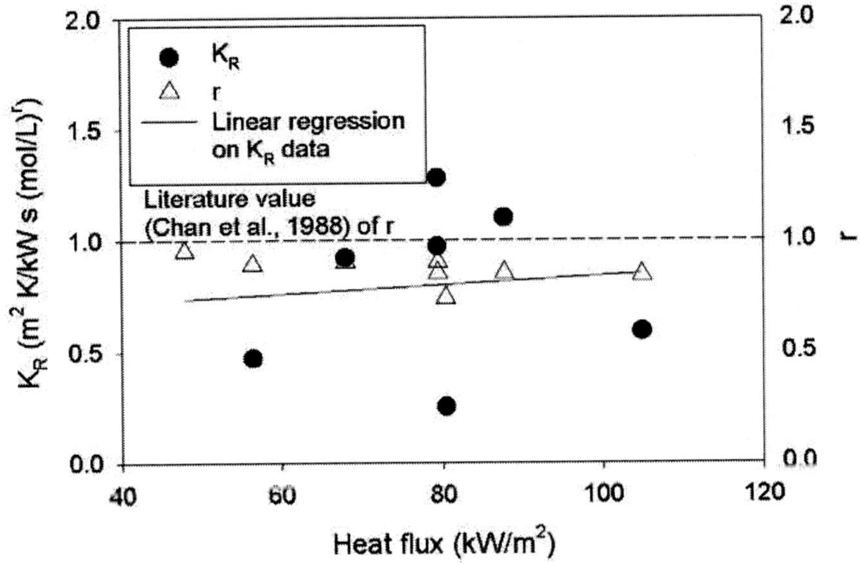

Fig. 6.2 The Chan et al. model parameters in comparison to literature value (K_R data for Run 4 not included)

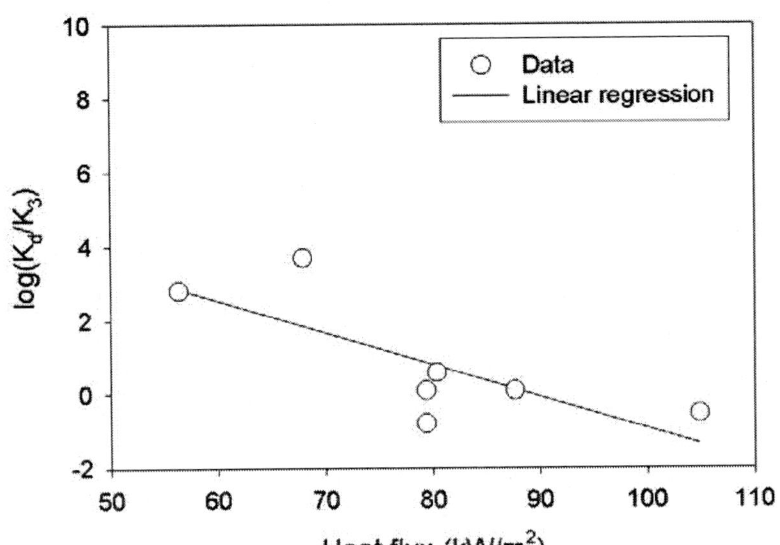

Fig. 6.3 The Watkinson and Epstein model parameters (data for Run 4 not available)

Fig 6.4 shows that experimental data were best predicted by the proposed composite fouling model for all conditions with a mean deviation of less than 6%, which was a similar magnitude to the experimental error (see section 3.2.3). The mean deviations of data fits for the Watkinson model and Chan's silica deposition model ranged from 25-35%. Other particulate fouling models from literature have also been tested with similar results to those of the above two models (e.g., the Epstein model and the Taborek et al. model with ca. 30-35% deviation). The differences in the model fits demonstrated the advantages of composite fouling model in describing COM-SiO_2 composite fouling behaviour by accounting for the interactive effects between CaOx and SiO_2 under subcooled boiling, which resulted in deposit consolidation, and bulk subcooling effect. Therefore the poor data fits by the other models were not surprising, as they were not derived to specifically involve these effects, which were critical to the reliable analysis of composite fouling process.

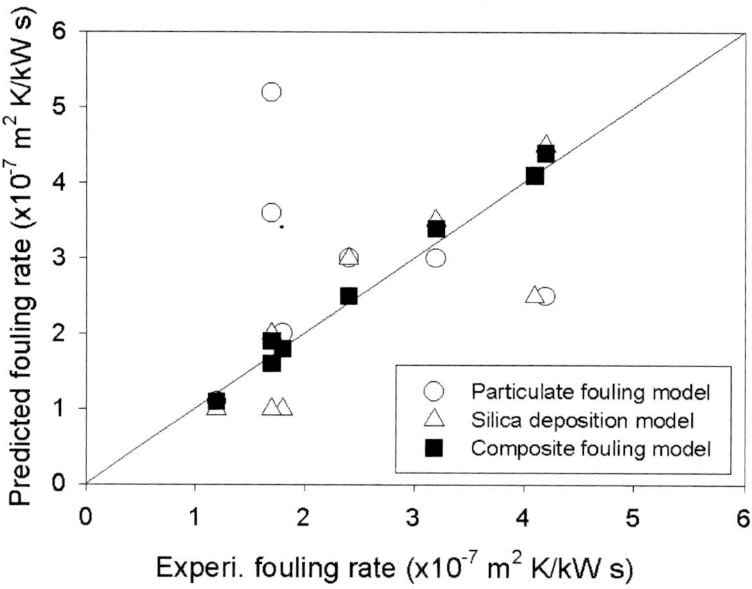

Fig. 6.4 Comparison of experimental data with model predictions

A major feature of the model presented in this study in contrast to previous particulate fouling models was that the effects of thermal hydraulic parameters on the rate of deposit consolidation during fouling process was considered based on the mechanism of microlayer evaporation, which was directly proportional to heat flux as was demonstrated in Asakura et al. model [eqn (6.17)]. Since heat flux increased with increasing fluid velocity and bulk subcooling, the higher velocity and subcooling also indirectly contributed to a higher degree of microlayer evaporation, and hence deposit consolidation (section 4.7).

Liquid bulk subcooling was one of the key elements in composite fouling. Apart from the role of changing the rate of deposit consolidation, there were other ways in which subcooling could affect the model parameters. Firstly, the physical properties of solution such as viscosity were dependent on bulk temperature. This included higher solution viscosity with increasing subcooling, which in turn affected the hydrodynamic lift forces at wall and removal rate of the deposit. Secondly, colloidal COM-SiO_2-sugar species (S) acquired their surface charges through the dissociation of protons from hydroxyl groups (see Fig 4.2.5 and Table 5.3.3), i.e.,

$$-SOH \overset{K_a}{\Leftrightarrow} -SO^- + H^+ \tag{6.27}$$

$$-SOH_2^+ \overset{K_b}{\Leftrightarrow} -SOH + H^+ \tag{6.28}$$

where K_a and K_b are the dissociation constants of eqn 6.27 and 6.28, respectively. The charge or the degree of dissociation of these species will be governed by the dissociation constants, which is always pH and temperature dependent. An increase in the point of zero charge (pzc) [pzc = $(pK_a+pK_b)/2$] of colloidal species could be obtained by decreasing the bulk temperature if the above equilibria were endothermic (Akratopulu et al., 1990), since fewer hydroxyl groups became negatively charged due to proton release. While the surface charge of the heat exchange tube remained unchanged at constant surface temperature, a reduction in the surface charge of the particle would affect the particle attachment efficiency by reducing the energy barrier at particle – heat exchange tube interface. The former (viscosity) effect has been accounted for by the use of higher velocity dependency of removal rate based on previous work by Williams et al. (1996), whereas the latter (surface charge effect) was not included in the model as it was unlikely that this constituted any significant contribution towards model parameters given the limited range of subcooling employed in the present study (see Table 6.3). It should be noted however, that the actual surface charge variation could be much more complex than that portrayed here, because of the wide variations in solution pH and the concentrations of sugar, CaOx and SiO_2, which might also have significant impact on the deposit removal and consolidation processes. While this is beyond the scope of the present study, further work to verify and incorporate these effects into the existing or other new mechanistic model are recommended to provide a better prediction of the composite fouling process in sugar mill evaporators.

Table 6.3 Surface zeta-potential of sample particles obtained at different subcooling levels

Bulk subcooling (K)	Surface zeta-potential of particle (mV)
20	-26.3*
25	-25.7
30	-24.9

* average of 6 runs

6.4 CONCLUSION

A new composite fouling model, which takes into account the additional effects of deposit consolidation and bulk subcooling, has been developed in this work. This new model attempts to describe the COM-SiO_2 composite fouling in an evaporator unit. The rate of fouling was better predicted by the proposed model in comparison to other fouling models over a range of thermal hydraulic parameters under subcooled nucleate flow boiling condition. However, the effects of solution pH and composition on the overall composite fouling process could not be determined from the new model further experiments and model refinement would be required in the future to establish the contribution of these parameters.

Chapter 7

CONCLUSIONS AND RECOMMENDATIONS

7.1 CONCLUSIONS

Coprecipitation of CaOx and SiO_2 has been shown to affect both CaOx precipitation and SiO_2 polymerization/precipitation in water and sugar solutions. The presence of SiO_2 in supersaturated solutions of CaOx led to a decrease in the rate of CaOx precipitation and an increase in the solubility of CaOx. The presence of CaOx in supersaturated solutions of SiO_2 enhanced the SiO_2 polymerization and lowered the SS ratio required for colloidal SiO_2 to form as a result of the formation of polymeric CaOx-SiO_2 complexes.

The presence of sugar accelerated both the precipitation of CaOx and polymerization of SiO_2 and reduced their solubilities in single foulant solutions. The threshold initial SiO_2 supersaturation required for the formation of colloidal SiO_2 species was slightly lowered by the presence of sugar (from 2.0 to 1.8). In the binary foulant sugar solutions, the presence of SiO_2 or sugar enhanced the precipitation of CaOx and reduced its solubility. CaOx or sugar either promoted or inhibited SiO_2 polymerization, depending on the initial CaOx and sugar concentrations. The presence of CaOx further decreased the threshold SiO_2 supersaturation required for colloidal SiO_2 formation (from 1.8 to 1.3), but had little effect on the solubility of SiO_2. These findings may be attributed to a number of mechanisms including sucrose hydration, adduct formation between CaOx, SiO_2 and sucrose, and adsorption of CaOx-sucrose complexes by polymeric SiO_2-sucrose species. Addition of SiO_2 or sugar in the CaOx crystallization system selectively promoted the formation of calcium oxalate trihydrate (COT) and calcium oxalate dihydrate (COD), respectively, and modified the crystal size and habit of calcium oxalate monohydrate (COM) crystals, the extent of which was dependant on the concentrations of SiO_2 and sugar. CaOx - SiO_2 co-precipitation in sugar solutions led to the stabilization of either COD or COT, depending on the initial concentration of CaOx. A variety of mechanisms including electrostatic effects, geometrical matching and stereochemical compatibility between CaOx crystallites and polymeric SiO_2/sucrose molecules have been proposed to account for the observed crystallographic and morphological changes of CaOx.

Based on a novel dynamic system designed to simulate the operation cycle of sugar mill evaporators, the composite fouling behaviour of COM and SiO_2 was tested in water and sugar solutions with a range of COM/SiO_2 supersaturation ratios to determine the critical feed composition at which the maximum degree of composite fouling occurred. The results showed that a synergistic effect of COM on composite fouling occurred at an intermediate SS of COM (2.6), whereas antagonistic effect was obtained at either low or high COM SS (1.1 and 5.3). The composite scale obtained from COM SS of 2.6 was of higher tenacity than that of the pure SiO_2 scale due to the cementing effect of COM on the deposit structure.

The primary mechanism of composite fouling was particulate deposition of the COM-SiO_2 colloidal species. The active interfacial forces involved are believed to be van der Waals forces of attraction and the electrical double layer repulsive forces. The rate of composite fouling was found to increase with decreasing interfacial energy barrier between the surface of heat exchange tube and COM-SiO_2 colloidal particle, which was determined by zeta-potential analysis, as a result of the interactions of the above forces. Thus, composite fouling was likely to be a surface-controlled process.

In addition to the role of interfacial forces in controlling particle deposition rate, the characteristics of subcooled boiling heat transfer and composite fouling were also strongly dependent on various thermal hydraulic parameters including surface superheat, fluid velocity and bulk subcooling. Firstly, the variations in the effects of heat flux, fluid velocity and subcooling on the observed heat transfer coefficients over a range of sugar concentrations implied an enhanced contribution of nucleate boiling heat transfer in the partial flow boiling regime, where both forced convection and nucleate boiling heat transfer occurred. The flow boiling heat transfer data could be predicted reasonably well by a modified Chen model. Secondly, the composite fouling rate increased with increasing superheat due to enhanced bubble-particle interactions at the boiling surface. For a constant superheat, the fouling rate reached a maximum at an intermediate subcooling but remained relatively constant over a range of fluid velocities. The composite deposit was found to exhibit higher bonding strength than silica scales due to structural consolidation of COM in the silica matrix by micro-layer evaporation under subcooled nucleate boiling. This led to linear rather than asymptotic fouling kinetics during composite fouling. The extent of deposit consolidation was shown to increase with superheat, fluid velocity and subcooling.

Composite scale formation of COM and SiO_2 from sugar solutions was described using a new analytical model, which incorporated the effects of liquid bulk subcooling and deposition consolidation resulting from microlayer evaporation under subcooled nucleate flow boiling. The proposed new composite fouling model provided a better fit to the COM-SiO_2 composite fouling data than models that considered only particulate fouling or silica deposition over a range of process thermal hydraulic parameters. Only two constants were necessary for each set of operating condition in order to predict the composite fouling rate in an evaporator unit of constant solution composition by the proposed model.

An implication of the results generated in this project is that the variability in scaling rates that occurs during the season and between seasons may well be dependent on the COM and SiO_2 concentrations and their ratios.

7.2 Recommendations

The following strategies for controlling the COM-SiO_2 composite fouling process are proposed:

- Fouling rate and scale tenacity may be minimized by avoiding operation under a critical feed composition, i.e., by adding either oxalic acid or SiO_2 in juice to alter the CaOx/SiO_2 SS ratio to the optimum value. This may be achieved by developing a device that operates in a sequential mode to measure and control the oxalic acid and SiO_2 concentrations in juice;
- The tenacity of composite scales may be reduced by controlling the thermal hydraulic operating conditions, especially to maintain the fluid velocity under a certain critical value (e.g., 1.1 m/s). This may be achieved by controlling the level of juice in the calandia tubes during operation and optimized evaoprator design (e.g., calandia geometry).

Future research in the following directions is recommended:

- The kinetic modeling of batch co-precipitation and identification of the chemical structure of COM-SiO_2-sucrose complexes, so that a better understanding of the COM-SiO_2 co-precipitation mechanisms can be obtained;
- Studies on the effect of solution composition on composite fouling rate and deposit characteristics;
- Further improvement of the composite fouling model by including the effects of pH and solution composition on the çomposite fouling process;
- The composite fouling behaviour of real factory juice needs to be evaluated before a better understanding of composite fouling mechanisms and appropriate methods of scale control can be established.

REFERENCES

Adomeit, P. and Renz, U., (2000). Correlations for the Particle Deposition Rate Accounting for Lift Forces and Hydrodynamic Mobility Reduction. *The Canadian Journal of Chemical Engineering*, 78: 32-39.

Akratopulu, K. C., Kordulis, C., and Lycourghiotis, A., (1990). Effect of Temperature on the Point of Zero Charge and Surface Charge of TiO_2, *Journal of Chemical Society Faraday Trans.*, 86: 3437-3440.

Alexander, G.B., Heston, W.M. and Iler, R.K., (1953). The Solubility of Amorphous Silica in Water. *JACS*, 75: 5655.

Andritsos, N. and Karabelas, A.J., (1999). The Influence of Particulates on $CaCO_3$ Scale Formation. *Journal of Heat Transfer*, 121: 225-227.

APHA, (1996). *Standard methods for the examination of water and wastewater.* APHA-AWWA-WEF, Washington, D.C.

Arnott, H.J., (1982). Three Systems of Biomineralization in Plants with Comments on the Associated Organic Matrix. In: G.H. Nancollas (Editor), *Biological Mineralization and Demineralization*. Springer-Verlag, pp. 199-218.

Asakura, Y., Kikuchi, M., Uchida, S. and Yusa, H., (1978). Deposition of Iron Oxide on Heated Surfaces in Boiling Water. *Nuclear Sci. Eng.*, 67: 1-7.

Atkins, P.W., (1994). *Physical Chemistry.* Oxford University Press, Oxford ; New York.

Babic-Ivancic, V., Furedi-Milhofer, H., Purgaric, B. and Brnicevic, N., (1985). Precipitation of Calcium Oxalates from High Ionic Strength Solutions III. The Influence of Reactant Concentrations on the Properties of the Precipitates. *Journal of Crystal Growth*, 71: 655-663.

Bansal, B., Muller-Steinhagen, H. and Chen, X.D., (1997). Effect of Suspended Particles on Crystallization Fouling in Plate Heat Exchangers. *Journal of Heat Transfer*, 119: 568-674.

Bao, Z.Y., Fletcher, D.F. and Haynes, B.S., (2000). Flow Boiling Heat Transfer of Freon R11 and HCFC 123 in Narrow Passages. *Int. Journal Heat Mass Transfer*, 43: 3347.

Basset, M., McInerney, J., Arbeau, N. and Lister, D.H., (2000). The Fouling of Alloy-800 Heat Exchange Surfaces by Magnetite Particles. *The Canadian Journal of Chemical Engineering*, 78: 40-52.

Baudin, I., Ricard, A. and Audebert, R., (1990). Adsorption of Dextrans and Pullulans at the Silica-Water Interface; Hydrodynamic Layer Thickness Measurements; Role in the

Fouling of Ultrafiltration Membranes. *Journal of Colloid & Interface Science*, 138(2): 324-331.

Bauer, D. et al., (1999). Stability of Colloidal Silica, Sikron and Polystyrene Latex Influenced by the Adsorption of Polycations of Different Charge Density. *Colloids and Surfaces A: Physicochemical and Engineering Aspects*, 156: 291-305.

Beal, S.K., (1973). Turbulent Agglomeration of Suspensions. *Journal of Aerosol Science*, 3: 113-125.

Beal, S.K., (1978). Correlations for the Sticking Probability and Erosion of Particles. *Journal of Aerosol Science*, 9: 455-461.

Beal, S.K., (1983). Particulate Fouling of Heat Exchangers. In: R.W. Bryers, Cole, S.S. (Editor), *Fouling of Heat Exchanger Surfaces*. Eng. Foundation, White Haven, Penn., pp. 215-233.

Bellamy, L.J., (1975). *The Infra-red Spectra of Complex Molecules*. Chapman and Hall.

Benjamin, R.J. and Balakrishnan, A.R., (1997). Nucleation Site Density in Pool Boiling of Saturated Pure Liquids: Effect of Surface Microroughness and Surface and Liquid Physical Properties. *Experimental Thermal and Fluid Science*, 15: 32.

Bergles, A.E. and Rohsenow, W.M., (1964). The Determination of Forced-Convection Surface-Boiling Heat Transfer. *Journal of Heat Transfer*, 86(3): 365.

Bjorge, R.W., Hall, G.R. and Rohsenow, W.M., (1982). Correlation of Forced Convection Boiling Heat Transfer Data. *Int. Journal Heat Mass Transfer*, 25(6): 753.

Bohnet, M., (1987). Fouling of Heat Transfer Surfaces. *Chem. Eng. Technol.*, 10: 113-125.

Bott, T.R., (1995). *Fouling of Heat Exchangers*. Elsevier Science B.V., Amsterdam, 524 pp.

Bott, T.R., (1997). The Fouling of Heat Transfer Surfaces, *Proc. of Engineering Foundation Conf. on Understanding Heat Exhanger Fouling and Its Mitigation*, pp. 3-16.

Bott, T.R. and Gudmundsson, J.S., (1978). Rippled Silica Deposits in Heat Exchanger Tubes, *Proc. 6th Int. Heat Transfer Conf.*, pp. 373-378.

Bramson, D., Hasson, D. and Semiat, R., (1995). The Roles of Gas Bubbling, Wall Crystallization and Particulate Deposition in $CaSO_4$ Scale Formation. *Desalination*, 100: 105-113.

Brecevic, L., Skrtic, D. and Garside, J., (1986). Transformation of Calcium Oxalate Hydrates. *Journal of Crystal Growth*, 74: 399-408.

Bridgwater, J. and Loo, C.E., (1984). Removal of Crystalline Scale: Mechanisms and the Role of Thermal Stress, *1st UK Heat Transfer Conf. Symposium Series No. 86*, pp. 455-463.

Burriel-Marti, F., Ramirez-Munoz, J. and Fernandez-Caldas, E., (1953). Determination of Oxalate Ion and Calcium Ion by Indirect Colorimetry. *Analytical Chemistry*, 25: 583-585.

Burrill, K.A., (1979). Corrosion Product Transport in Water-cooled Nuclear Reactors, Part III: Boiling Water with Indirect Cycle Operation. *The Canadian Journal of Chemical Engineering*, 57: 211-224.

Burton, W.K., Cabrera, N. and Frank, F.C., (1951). The Growth of Crystals and the Equilibrium Structure of Their Surfaces. *Phil. Trans. Royal Soc.*, A243: 299-358.

Butterworth, D. and Shock, R.A.W., (1982). Flow Boiling, *Proc. 7th Int. Heat Transfer Conf.*, pp. 11.

Callejas-Fernandez, J., de las Nieves, F.J., Martinez-Garcia, R. and Hidalgo-Alvarez, R., (1991). Electrokinetic Characterization and Colloidal Stability of Calcium Oxalate

Monohydrate Dispersions in the Presence of Certain Inhibitors. *Progress in Colloid & Polymer Science*, 84: 327-333.

Callejas-Fernandez, J., Martinez-Garcia, R., de las Nieves Lopez, J. and Hidalgo-Alvarez, R., (1993). Electrosuperficial Properties of Calcium Oxalate Hydrates: Crystal Growth and Ionic Reactivity. *Solid State Ionics*, 63-65: 791-796.

Campbell, A.A., Ebrahimpour, A., Perez, L., Smesko, S.A. and Nancollas, G.H., (1989). The Dual Role of Polyelectrolytes and Proteins as Mineralization Promoters and Inhibitors of Calcium Oxalate Monohydrate. *Calcif. Tissue Int.*, 45: 122-128.

Campbell, N.A.F., Hawley, J.G., Robinson, K. and Leathard, M.J., (2000). A Review of Predictive Analysis Applied to IC Engine Coolant Heat Transfer. *Journal of the Institute of Energy*, 73: 78.

Carroll, B. and Freeman, E., (1954). The Behavior of Colloidal Silicate Solutions as Revealed by Adsorption Indicators. *JACS*: 335-338.

Celata, G.P., Cumo, M. and Setaro, T., (1993). Forced Convective Boiling in Binary Mixtures. *Int. Journal Heat Mass Transfer*, 36(13): 3299.

Chan, S.H., (1988). Fouling of Geothermal Energy Systems. In: L.F. Melo (Editor), *Fouling Sci. & Tech.* Kluwer Academic Publisher, Dordrecht, The Netherlands, pp. 649-667.

Chan, S.H., (1989). A Review on Solubility and Polymerization of Silica. *Geothermics*, 18: 49-56.

Chan, S.H., Rau, H., Debellis, C. and Neusen, K.F., (1988). Silica Fouling of Heat Transfer Equipment - Experiments and Model. *Journal of Heat Transfer*, 110: 841-849.

Chang, G.T. and Bankoff, S.G., (1990). On the Mechanism of Forced-Convection Subcooled Nucleate Boiling. *Journal of Heat Transfer*, 112: 213-218.

Charlesworth, D.H., (1970). The Deposition of Corrosion Products in Boiling Water Systems. *Chemical Engineering Progress Symposium Series*, 66: 21-30.

Chen, J.C., (1966). Correlation for Boiling Heat Transfer to Saturated Fluids in Convective Flow. *I&EC Process Design and Development*, 5(3): 322.

Chen, J.C. and Tuzla, K., (1995). Contribution of Convection and Boiling to Convective Flow Boiling. In: J.C. Chen (Editor), *Proceedings of Convective Flow Boiling*. Taylor & Francis, Alberta, Canada, pp. 181.

Cleaver, J.W. and Yates, B., (1975). A Sub-layer Model for the Deposition of Particles from a Turbulent Flow. *Chem. Eng. Sci.*, 30: 983-992.

Cody, A.M. and Cody, R.D., (1994). Calcium Oxalate Trihydrate Phase Control by Structurally-specific Carboxylic Acids. *Journal of Crystal Growth*, 135: 235-245.

Cooper, M.G., (1984). Saturation Nucleate Pool Boiling - A Simple Correlation, Proceedings of the First U.K. National Conference on Heat Transfer. *Inst. Chem. Eng. Symp.* Ser. no. 86, London, pp. 785.

Cordiner, J.B., *et al.*, (1969). Heat Transfer Coefficient for Raw Sugar Solutions. *The Sugar Journal*: 13.

Crawford, J.E., Crematy, E.P. and Alexander, A.E., (1968). The Effect of Natural and Synthetic Polyelectrolytes on the Crystallization of Calcium Oxalate. *Aust. J. Chem.*, 21: 1067-1072.

Deganello, S., Kampf, A.R. and Moore, P.B., (1981). The Crystal Structure of Calcium Oxalate Trihydrate: $Ca(H_2O)_3(C_2O_4)$. *American mineralogist*, 66: 859-865.

Dhir, V.K., (1998). Boiling Heat Transfer. *Annu. Rev. Fluid Mech.*, 30: 365.

Dirksen, J.A. and Ring, T.A., (1991). Fundamentals of Crystallization Kinetic Effects on Particle Size Distributions and Morphology. *Chem. Eng. Sci.*, 46(10): 2389-2427.

Doherty, W.O.S., (2000). Chemical Cleaning of Sugar Mill Evaporator. *Proc. Aust. Soc. Sugar Cane Technol.*, 22: 341-346.

Doherty, W.O.S., (2002-2004). *Personal Communications.*

Doherty, W.O.S., Crees, O.L. and Senogles, E., (1994). The Precipitation of Calcium Oxalate Dihydrate Crystals. *Crystal Research Technology*, 29(4): 517-524.

Doherty, W.O.S., Crees, O.L. and Senogles, E., (1995). Polymeric Additives' Effect on Crystallization of Calcium Oxalate Scales. *Crystal Research Technology*, 30: 791-800.

Eggleston, G. and Vercellotti, J.R., (2000). Degradation of Sucrose, Glucose and Fructose in Concentrated Aqueous Solutions under Constant pH Conditions at Elevated Temperature. *Journal of Carbohydrate Chemistry*, 19(9): 1305-1318.

Epstein, N., (1981a). Fouling in Heat Exchangers. In: E.F.C. Somerscales (Editor), *Fouling of Heat Transfer Equipments*. Hemisphere Publishing Co., pp. 701-731.

Epstein, N., (1981b). Fouling: Technical Aspects (Afterword to Fouling in Heat Exchangers). In: E.F.C. Somerscales (Editor), *Fouling of Heat Transfer Equipments*. Hemisphere Publishing Co., pp. 31-51.

Epstein, N., (1983). Thinking about Heat Transfer Fouling: A 5 x 5 Matrix. *Heat Transfer Engineering*, 4(1): 43-56.

Epstein, N., (1988). Particulate Fouling of Heat Transfer Surfaces: Mechanisms and Models. In: L.F. Melo, *et al.* (Editor), *Fouling Science and Technology*. Kluwer Academic Publishers, Dordrecht, The Netherlands, pp. 143-164.

Epstein, N., (1997). Elements of Particle Deposition onto Nonporous Solid Surfaces Paralel to Suspension Flows. *Experimental Thermal and Fluid Science*, 14: 323-334.

Epstein, N., (1999). Particle Deposition and Its Mitigation. In: T.R. Bott (Editor), *Proceedings of an International Conference on Understanding Heat Exchanger Fouling and Its Mitigation.* Begell House Inc., Castelvecchio Pascoli, Italy, pp. 3-22.

Esumi, K. and Matsui, H., (1993). Adsorption of Non-Ionic and Cationic Polymers on Silica from Their Mixed Aqueous Solutions. *Colloids and Surfaces A: Physicochemical and Engineering Aspects*, 80: 273-278.

Forster, H.K. and Zuber, N., (1955). Dynamics of Vapor Bubbles and Boiling Heat Transfer. *AICHE Journal*, 1(4): 531.

Franceschi, V.R. and Horney, H.T., Jr., (1980). Calcium Oxalate Crystals in Plants. *The Botanical Review*, 46(4): 361-427.

Freeborn, J. and Lewis, D., (1962). Initiation of Boiler Scale Formation. *Journal of Mechanical Engineering Science*, 4(1): 46-52.

Fujita, Y., (1992). The State of the Art - Nucleate Boiling Mechanisms. *Pool and External Flow Boiling*: 83.

Furedi-Milhofer, H. et al., (1990). Factors Influencing Nucleation from Solutions Supersaturated to Different Crystal Hydrates. *Colloids and Surfaces*, 48: 219-230.

Furedi-Milhofer, H., Markovic, M. and Uzelac, M., (1987). Precipitation and Solubility of Calcium Phosphates and Oxalates in the System $Ca(OH)_2$-H_3PO_4-$H_2C_2O_4$-NaCl-H_2O. *Journal of Crystal Growth*, 80: 60-68.

Furedi-Milhofer, H., Sikiric, M., Tunik, L., Filipovic-Vincekovic, N. and Garti, N., (2002). Interactions of Organic Additives with Ionic Crystal Hydrates: the Importance of the Hydrated Layer. *International Journal of Modern Physics B*, 16(1-2): 359-366.

Garrett-Price, B.A. et al., (1985). *Fouling of Heat Exchangers*. Noyes Publications, Park Ridge, NJ, 417 pp.

Garside, J., (1982). Nucleation. In: G.H. Nancollas (Editor), *Biological Mineralization and Demineralization*. Springer-Verlag, pp. 23-36.

Garside, J., Brecevic, L. and Mullin, J.W., (1982). The Effect of Temperature on the Precipitation of Calcium Oxalate. *Journal of Crystal Growth*, 57: 233-240.

Garyazha, V.T. and Kulinchenko, V.R., (1975). Investigation of the Mechanism of Boiling Heat Transfer to Water and Sugar Solutions. *Heat Transfer - Soviet Research*, 7(5): 12.

Gill, J.S., (1998). Silica Scale Control. *Materials Performance*: 41-45.

Grases, F., Millan, A. and Conte, A., (1990). Production of Calcium Oxalate Monohydrate, Dihydrate or Trihydrate. *Urol. Res.*, 18: 17-20.

Gregory, J., (1989). Fundamentals of Flocculation. *Critical Reviews in Environmental Control*, 19: 185-230.

Gudmundsson, J.S., 1979. *Fouling of Surfaces*. Ph.D Thesis, University of Birmingham, Birmingham.

Gudmundsson, J.S., (1981). Particulate Fouling. In: E.F.C. Somerscales (Editor), *Fouling of Heat Transfer Equipment*, pp. 357-387.

Guglielmini, G., Nannei, E. and Pisoni, C., (1988). Survey of Heat Transfer Correlations in Forced Convection Boiling. In: T.N. Veziroglu (Editor), *Multiphase Transport*. Hemisphere Publishing Co., pp. 845.

Gungor, K.E. and Winterton, R.H.S., (1986). A General Correlation for Flow Boiling in Tubes and Annuli. *Int. Journal Heat Mass Transfer*, 29(3): 351.

Heijnen, W.M.M., (1982). The Growth Morphology of Calcium Oxalate Trihydrate - A contribution to Urinary Stone Research. *Journal of Crystal Growth*, 57: 216-232.

Helalizadeh, A., Muller-Stainhagen, H. and Jamialahmadi, M., (2005). Mathematical Modelling of Mixed Salt Precipitation during Convective Heat Transfer and Sub-cooled Flow Boiling. *Chemical Engineering Science*, 60: 5078-5088.

Henley, M., (1992). Colloidal and Soluble Silica Removal Remains Challenge When Producing High-purity Water. *Ultrapure Water*, 9(9): 13.

Hess, B., Nakagawa, Y., Parks, J.H. and Coe, F.L., (1991). Molecular Abnormality of Tamm-Horsfall Glycoprotein in Calcium Oxalate nephrolithiasis. *American Journal of Physiology*, 260: F569-578.

Hong, S. and Elimelech, M., (1997). Chemical and Physical Aspects of Natural Organic Matter (NOM) Fouling of Nanofiltration membranes. *Journal of Membrane Science*, 132: 159-181.

Honig, P., (1963). *Principle of Sugar Technology, III*. Elsevier Publishing Co., 711 pp.

Hospeti, N.B. and Mesler, R.B., (1965). Deposits Formed Beneath Bubbles During Nucleate Boiling of Radioactive Calcium Sulfate Solutions. *AICHE Journal*, 11(4): 662-665.

Hussain, C.I., Bott, T.R. and Newson, I.H., (1986). Heat Exchanger Fouling: A Study of Underlying Mechanisms for Particulate Deposition. *IChemE Symposium Series*, 86: 369-379.

Iler, R.K., (1975). Coagulation of Colloidal Silica by Calcium Ions, Mechanism, and Effect of Particle Size. *Journal of Colloid & Interface Science*, 53(3): 476-488.

Iler, R.K., (1979). *The Chemistry of Silica - Solubility, Polymerization, Colloid and Surface Properties, and Biochemistry*. John Wiley & Sons, Inc., 866 pp.

Israelachvili, J.N., (1991). *Intermolecular and Surface Forces.* Academic Press, New York, 450 pp.

Jamialahmadi, M. and Muller-Steinhagen, H., (1993). Effect of Process Parameters on Scale Formation from Spent Bayer Process Liquor - Experimental Observations, Part II. *Aluminium*, 69(10): 920-923.

Kaftori, D., Hetsroni, G. and Banerjee, S., (1995). Particle Behaviour in the Turbulent Boundary Layer. I. Motion, Deposition, and Entrainment. *Phys. Fluids*, 7(5): 1095-1106.

Kandlikar, S.G., (1983). An Improved Correlation for Predicting Two-Phase Flow Boiling Heat Transfer Coefficient in Horizontal and Vertical Tubes. In: J.B.J. Kitto (Editor), *The 21st National Heat Transfer Conf.* ASME, Seattle, Wash., pp. 3.

Kern, D.Q. and Seaton, R.E., (1959). Theoretical Analysis of Thermal Surface Fouling. *Brit. Chem. Eng.*, 4(5): 258-262.

Kitano, Y., Okumura, M. and Idogaki, M., (1979). Behavior of Dissolved Silica in Parent Solution at the Formation of Calcium Carbonate. *Geochemical Journal*, 13: 253-260.

Klimas, S.J., Pietralik, J.M., Fruzzetti, K. and Tapping, R.L., (2004). Fouling Enhancement under Flow Boiling at Elevated Steam Qualities. *International Journal of Transport Phenomena*, 6(2).

Knudsen, J.G., (1981). Cooling Water Fouling - A Brief Review, Fouling in Heat Transfer Equipment, *20th ASME/AICHE Heat Transfer Conference*, Milwaukee, pp. 29-38.

Knudsen, J.G. and Roy, B.V., (1982). Influence of Fouling on Heat Transfer, *7th Int. Heat Transfer Conf.*, Munich, pp. 269-287.

Kollonitsch, V., (1970). *Sucrose Chemicals.* The International Sugar Research Foundation, Inc.

Kontoyannis, C.G., Bouropoulos, N.C. and Koutsoukos, P.G., (1997). Urinary Stone Layer Analysis of Mineral Components by Raman Spectroscopy, IR Spectroscopy, and X-ray Powder Diffraction: A Comparative Study. *Applied Spectroscopy*, 51(8): 1205-1209.

Kreith, F., (1973). *Principles of Heat Transfer.* Intext Press, Inc., New York, NY, 656 pp.

Krumbein, W.E. and Werner, D., (1983). The Microbial Silica Cycle. In: W.E. Krumbein (Editor), *Microbial Geochemistry.* Blackwell Scientific Fublications, pp. 330.

Lavarack, B.P., (2001). Reducing Process Steam Usage? What Will Happen to the Water Balance? *Proc. Aust. Soc. Sugar Cane Technol.*, 23: 410-415.

Leenaars, A.F.M., (1988). A New Approach to the Removal of Sub-micron Particles from Solid (Silicon) Substrates. In: K.L. Mittal (Editor), *Particles on Surfaces: Detection, Adhesion, and Removal.* Plenum, New York, pp. 361.

Lencar, D. and Watkinson, A.P., (2001). *Fouling by Calcium Oxalates in Aqueous Solutions, Heat Transfer Fouling Fundamental Approaches & Technical Solutions.* United Engineering Foundation, Inc., Davos, Switzerland.

Letellier, S.R., Lochhead, M.J., Campbell, A.A. and Vogel, V., (1998). Oriental Growth of Calcium Oxalate Monohydrate Crystals beneath Phospholipid Monolayers. *Biochimica et Biophysica Acta*, 1380: 31-45.

Levy, B. and Fritsch, A.R., (1959). Electrokinetic Measurements on Stainless Steel Capillaries. *Journal of the Electrochemical Society*, 106(8): 730-735.

Lister, D.H., (1981). Corrosion Fouling and Fouling due to Liquid Solidification. In: E.F.C. Somerscales, Knudsen, J.G. (Editor), *Fouling of Heat Transfer Equipment.* Hemisphere Pub. Corp., pp. 136-200.

Liu, Z. and Winterton, R.H.S., (1991). A General Correlation for Saturated and Subcooled Flow Boiling in Tubes and Annuli, Based on a Nucleate Pool Boiling Equation. *Int. Journal Heat Mass Transfer*, 34(11): 2759.

Long, R.B., (1995). *Separation Processes in Waste Minimization*. Marcel Dekker, Inc., 472 pp.

Mann, S., (1988). Molecular Recognition in Biomineralization. *Nature*, 332(10): 119-124.

Manne, J.S., Biala, N., Smith, A.D. and Gryte, C.C., (1990). The Effect of Anionic Polyelectrolytes on the Crystallization of Calcium Oxalate Hydrates. *Journal of Crystal Growth*, 100: 627-634.

Mantel, M., Rabinovich, Y.I., Wightman, J.P. and Yoon, R.H., (1995). A study of hydrophobic interactions between stainless steel and silanated glass surface using atomic force microscopy. *Journal of Colloid & Interface Science*, 170: 203-214.

Mayinger, F. and Hollborn, E., (1977). The Effect of Liquid Viscosity on Bubble Formation and Heat Transfer in Boiling. In: E. Hahne, Grigull, U. (Editor), *Heat Transfer in Boiling*. Academic Press, pp. 391.

McGarvey, G.B. and Turner, C.W., (1999). Synergism Models for Heat Exchanger Fouling Mitigation. In: T.R. Bott (Editor), *Int. Conference on Understanding Heat Exchanger Fouling and Its Mitigation*. Begell House, Inc., Castelvecchio Pascoli, Italy.

Melia, T.P., (1965). Crystal Nucleation from Aqueous Solution. *Journal of Applied Chemistry*, 15: 345-357.

Melo, L.F., (1988). Particulate Fouling in Boiling Water Systems. In: S. Kakac, Bergles, A.E., Fernandes, E.O. (Editor), *Two-Phase Flow Heat Exchangers*. Kluwer Academic Publishers, pp. 461-469.

Melo, L.F. and Pinheiro, J.D., (1986). Particulate Fouling: Controlling Processes and Deposit Structure, *Proc. 8th Int. Heat Transfer Conf.*, pp. 2781-2786.

Metzner, A.B. and Friend, W.L., (1958). Theoretical Analogies between Heat, Mass and Momentum Transfer and Modifications for Fluids of High Prandtl or Schmidt Numbers. *Can. J. Chem. Eng.*, 36(6): 235-240.

Millan, A., (2001). Crystal Growth Shape of Whewellite Polymorphs: Influence of Structure Distortions on Crystal Shape. *Crystal Growth and Design*, 1(3): 245-254.

Millan, A., Pavelkova, M., Sohnel, O. and Grases, F., (1998). Inhibition of Calcium Oxalate Monohydrate Crystal Growth in High and Low Ionic Strength Solutions. *Crystal Research Technology*, 33(5): 777-786.

Moles, F.D. and Shaw, J.F.G., (1972). Boiling Heat Transfer to Subcooled Liquids under Conditions of Forced Convection. *Trans. Inst. Chem. Eng.*, 50: 76.

Morison, K., (2002). Viscosity Equation for Sucrose Solutions: Old and New, *Proceedings of 9th APCChE Congress and CHEMECA '02*, Christchurch, New Zealand, pp. CD-ROM paper 984.

Muller-Steinhagen, H., (1999). Cooling-Water Fouling in Heat Exchangers. *Advances in Heat Transfer*, 33: 415-496.

Muller-Steinhagen, H., Epstein, N. and Watkinson, A.P., (1986). Subcooled-Boiling and Convective Heat Transfer for Heptane Flowing Inside an Annulus and Past a Coiled Wire: Part II - Correlation of Data. *Journal of Heat Transfer*, 108: 928.

Muller-Stainhagen, H., Malayeri, M.R. and Watkinson, A.P., (2005). Fouling of Heat Exchangers - New Approaches to Solve an Old Problem. *Heat Transfer Engineering*, 26(1): 1-4.

Mullin, J.W., (1979). Crystallization, *Kirk-Othmer Encyclopedia of Chemical Technology.* John Wiley & Sons, Inc., New York, pp. 244-285.

Mullin, J.W., (1997). *Crystallization.* Butterworth-Heinermann, Oxford, UK, 527 pp.

Myerson, A.S., (1993). *Handbook of Industrial Crystallization.* Butterworth-Heinemann, Boston, 242 pp.

Nakamura, T., Sameshima, K., Okunaga, K., Sugiura, Y. and Sato, J., (1989). Determination of Amorphous Phase in Quartz Powder by X-ray Powder Diffractometry. *Powder Diffraction*, 4(1): 9-13.

Nancollas, G.H., (1982). Phase Transformation during Precipitation of Calcium Salts. In: G.H. Nancollas (Editor), *Biological Mineralization and Demineralization.* Springer-Verlag, pp. 79-99.

Nancollas, G.H., (1983). The Mechanisms of Precipitation of Biological Minerals. The Phosphates, Oxalates and Carbonates of Calcium. *Croatica Chemica Acta*, 56: 741-752.

Nancollas, G.H. and Gardner, G.L., (1974). Kinetics of Crystal Growth of Calcium Oxalate Monohydrate. *Journal of Crystal Growth*, 21: 267-276.

Nancollas, G.H. et al., (1991). Physical Chemical Studies of Calcium Oxalate Crystallization. *American Journal of Kidney Diseases*, 17(4): 392-395.

Nandagopal, S. and Ramamurthy, R., (1976). Evaporator - Scale Formation. *Indian Sugar*: 437-439.

Nicholson, F.D. and Sarbutt, J.V., (1980). The Effect of Boiling on the Mass Transfer of Corrosion Products in High Temperature, High Pressure Water Circuits. *Corrosion*, 36(1): 1-9.

Nielsen, A.E., (1960). The Kinetics of Calcium Oxalate Precipitation. *Acta Chemica Scandinavica*, 14: 1654-1659.

Ning, R.Y., (2002). Discussion of Silica Speciation, Fouling, Control and Maximum Reduction. *Desalination*, 151: 67-73.

Nishikawa, K. and Fujita, Y., (1976). On the Nucleation Factor in Nucleate Boiling Heat Transfer. *Memoirs of the Faculty of Engineering, Kyushu University*, 36(2): 139.

Nishino, Y., Asakura, Y., Sawa, T. and Uchida, S., (1991). Deposition of Nickel and Cobalt Ions with Iron Crud on Heated Stainless Steel Surface under Nucleate Boiling Conditions. *Journal of Nuclear Science and Technology*, 28(9): 848-857.

O'Callaghan, M.G., (1986). Fouling of Heat Transfer Equipment: Summary Review. In: J.W. Palen (Editor), *Heat Exchanger Sourcebook.* Hemisphere Publishing Co., pp. 745-755.

Ohtaki, H., (1998). *Crystallization Processes.* John Wiley and Sons Ltd., 208 pp.

Oliveira, R., Melo, L. and Pinheiro, M., (1988). Fouling by Aqueous Suspensions of Kaolin and Magnetite, *Proceedings of the 2nd UK National Conference on Heat Transfer. Mech. Engineering Publications,* Glasgow, pp. 161-171.

Oliveira, R., Melo, L., Pinheiro, M. and Vieira, M.J., (1993). Surface Interactions and Deposit Growth in Fouling of Heat Exchangers. *Corrosion Reviews*, 11(1-2): 55-95.

Pannetier, N., Khoukh, A. and Francois, J., (2001). Physico-chemical Study of Sucrose and Calcium Ions Interactions in Alkaline Aqueous Solutions. *Macromol. Symp.*, 166: 203-208.

Papavergos, P.G. and Hedley, A.B., (1984). Particle Deposition Behaviour from Turbulent Flows. *Chem. Eng. Res. Des.*, 62: 275-295.

Partridge, E.P. and White, A.H., (1929a). Mechanism of Formation of Calcium Sulfate Boiler Scale. *Industrial & Engineering Chemistry,* 21(9): 834-838.

Partridge, E.P. and White, A.H., (1929b). Thermal Effects of Boiler Scale. *Industrial & Engineering Chemistry*, 21(9): 839-844.

Patterson, R.E., (1993). *Silica, Kirk-Othmer encyclopedia of chemical technology*. John Wiley & Sons, Inc., pp. 982-1005.

Peacock, S.D., (1995). Predicting Physical Properties of Factory Juice and Syrups. *International Sugar Journal,* 97(1162): 571-577.

Peacock, S.D. and Starzak, M., (1996). Modelling of Climbing Film Evaporators. *Proc. South Afr. Sugar Technol. Ass.*: 213.

Perry, C.C. and Keeling-Tucker, T., (2000). Model Studies of the Precipitation of Silcia in the Presence of Aluminium; Implication for Biology and Industry. *Journal of Inorganic Biochemistry,* 78: 331-339.

Pinheiro, J.R.S., (1986). Fouling of Heat Transfer Surfaces. In: J.W. Palen (Editor), *Heat Exchanger Sourcebook*. Hemisphere Publishing Co., pp. 721-743.

Pioro, I.L., (1999). Experimental Evaluation of Constants for the Rohsenow Pool Boiling Correlation. *Int. Journal Heat Mass Transfer*, 42: 2003.

Poynton, M. and Alexander, J.B., (1957). Some Notes on the Solubility of Silica in Sugar Solutions. *The S.A. Sugar Journal*: 42-43.

Prodanovic, V., Fraser, D. and Salcudean, M., (2002). On the Transaction from Partial to Fully Developed Subcooled Flow Boiling. *Int. Journal Heat Mass Transfer*, 45: 4727.

Pulling, D.J. and Collier, J.G., (1963). The Deposition of solids in a Vertical Tube Evaporator. *The Industrial Chemist*, 39: 129-133.

Quraishi, M.S. and Varshney, B.S., (1975). Pool Boiling of Aqueous Sugar Solutions. *Indian Journal of Technology*, 13: 121.

Reeks, M.W. and Hall, D., (2001). Kinetic models for particle resuspension in turbulent flows: Theory and measurements. *Aerosol Science*, 32: 1-31.

Reichardt, C., (1988). *Solvents and Solvent Effects in Organic Chemistry*. VCH Verlagsgesellschaft mbH, Weinheim.

Robertson, W.G., (1982). The Solubility Concept. In: G.H. Nancollas (Editor), *Biological Mineralization and Demineralization*. Springer-Verlag, pp. 5-22.

Rohsenow, W.M., (1951). A Method of Correlating Heat-transfer Data for Surface Boiling of Liquids. *Trans. of the ASME*, 74: 969.

Rohsenow, W.M., (1953). *Heat Transfer with Evaporation, Heat Transfer*. University of Michigan Press, Ann Arbor, pp. 101.

Rohsenow, W.M., (1972). On the Rohsenow Pool-Boiling Correlation. *Journal of Heat Transfer*: 255.

Rothbaum, H.P. and Rohde, A.G., (1979). Kinetics of Silica Polymerization and Deposition from Dilute Solutions between 5 and 180°C. *Journal of Colloid & Interface Science*, 71: 533-559.

Ryall, R., et al. (Editor), (1994). *Urolithiasis 2*. International Symposium on Urolithiasis. Plenum Press, New York.

Sahai, N. and Tossell, J.A., (2000). ^{29}Si NMR Chemical Shifts and Energetics of Silica-Serine and Silica-Polyalcohol Complexes as Indicators of Silica Biomineralization Mechanisms. In: P.J. Li, Calvert, P., Kokubo, T., Levy, R., Scheid, C. (Editor), *Mineralization in Natural and Synthetic Biomaterials*. Materials Research Society, Warrendale, Penn., pp. 249-254.

Schwarz, S. et al., (2000). Adsorption and Stability of Colloidal Silica. *Colloids and Surfaces A: Physicochemical and Engineering Aspects*, 163: 17-27.

Shah, M.M., (1983). Generalized Prediction of Heat Transfer during Subcooled Boiling in Annuli. *Heat Transfer Engineering*, 4(1): 24.

Sheehan, M.E., (1981). *The Kinetics of Crystal Growth of Calcium Oxalate*. Ph.D Thesis, The State University of New York, 242 pp.

Sheikholeslami, R., Al-Mutaz, I.S., Koo, T. and Young, A., (2001). Pretreatment and the Effect of Cations and Anions on Prevention of Silica Fouling. *Desalination*, 139: 83-96.

Sheikholeslami, R. and Ng, M., (2001). Calcium Sulfate Precipitation in the Presence of Nondominant Calcium Carbonate: Thermodynamics and Kinetics. *Ind. Eng. Chem. Res.*, 40: 3570-3578.

Sheikholeslami, R. and Tan, S., (1999). Effects of Water Quality on Silica Fouling of Desalination Plants. *Desalination*, 126: 267-280.

Sikiric, M., Filipovic-Vincekovic, N., Babic-Ivancic, V., Vdovic, N. and Furedi-Milhofer, H., (1999). Interactions in Calcium Oxalate Hydrate/Surfactant Systems. *Journal of Colloid & Interface Science*, 212: 384-389.

Singh, R.P., Gaur, S.S., Sheehan, M.E. and Nancollas, G.H., (1988). Crystal Growth of Calcium Oxalate Monohydrate. *Journal of Crystal Growth*, 87: 318-324.

Sivagnanam, P., Balakrishnan, A.R. and Varma, Y.B.G., (1994). On the Mechanism of Subcooled Flow Boiling of Binary Mixtures. *Int. Journal Heat Mass Transfer*, 37(4): 681.

Sivagnanam, P. and Varma, Y.B.G., (1990). Subcooled Boiling of Binary Mixtures under Conditions of Forced Convection. *Experimental Thermal and Fluid Science*, 3: 515.

Skrtic, D., Furedi-Milhofer, H. and Markovic, M., (1987). Precipitation of Calcium Oxalates from High Ionic Strength Solutions. *Journal of Crystal Growth*, 80: 113-120.

Sohnel, O. and Garside, J., (1992). *Precipitation - Basic Principles and Industrial Applications*. Butterworth Heinemann, 390 pp.

Somerscales, E.F.C., (1992). The Effect of Fouling on Boiling Heat Transfer. In: R.T. Lahey, Jr. (Editor), *Boiling Heat Transfer: Modern Developments and Advances*. Elsevier Science B.V., Amsterdam, pp. 614.

Somerscales, E.F.C., (1999). Fundamentals of corrosion fouling. *British Corrosion Journal,* 34(2): 109-124.

Starzak, M. and Peacock, S.D., (2000). Hydration Number and Water Activity Models for the Sucrose-Water System: A Critical Review. *Critical Reviews in Food Science and Nutrition*, 40(4): 327-367.

Steiner, D. and Taborek, J., (1992). Flow Boiling Heat Transfer in Vertical Tubes Correlated by an Asymptotic Model. *Heat Transfer Engineering*, 13(2): 43.

Stephen, H. and Stephen, T., (1963). *Solubilities of Inorganic and Organic Compounds*, 1. Pergamon Press.

Streit, J., Tran-Ho, L. and Konigsberger, E., (1998). Solubility of the Three Calcium Oxalate Hydrates in Sodium Chloride Solutions and urine-Like Liquors. *Monatshefte fur Chemie*, 129: 1225-1236.

Sudmalis, M. and Sheikholeslami, R., (2000). Coprecipitation of $CaCO_{33}$ and $CaSO_4$. *The Canadian Journal of Chemical Engineering*, 78: 21-31.

Suitor, J.W., Marner, W.J. and Ritter, R.B., (1977). The History and Status of Research in Fouling of Heat Exchangers in Cooling Water Service. *The Canadian Journal of Chemical Engineering*, 55: 374-380.

Sutor, D.J., (1969). Growth Studies of Calcium Oxalate in the Presence of Various Ions and Compounds. *Brit. J. Urol.*, 41: 171-178.

Taborek, J., Aoki, T., Ritter, R.B. and Palen, J.W., (1972a). Predictive Methods for Fouling Behaviour. *Chemical Engineering Progress*, 68(7): 69-78.

Taborek, J., Aoki, T., Ritter, R.B., Palen, J.W. and Knudsen, J.G., (1972b). Fouling: The Major Unresolved Problem in Heat Transfer. *Chmical Engineering Progress*, 68(2): 59-67.

Tadros, T.F. and Lyklema, J., (1969). The Electrical Double Layer on Silica in the Presence of Bivalent Counterions. *ElectroAnalytical Chemistry & Interfacial Electrochemistry*, 22: 1-7.

Thom, J.R.S., Walker, W.M., Fallon, T.A. and Reising, G.F.S., (1965). Boiling in Subcooled Water during Flow Up Heated Tubes or Annuli. *Proc. Inst. Mech. Engrs.*, 180(Pt 3C): 226.

Tomazic, B. and Nancollas, G.H., (1980). Crystal Growth of Calcium Oxalate Hydrates: A Comparative Kinetics Study. *Journal of Colloid & Interface Science*, 75(1): 149-160.

Tomlinson, L., Ashmore, C.B., Silver, P.J.B. and Mead, A.P., (1990). The High Temperature Aqueous Corrosion of Ferritic Steels: Effect of Heat Flux on Corrosion and Magnetite Deposition. *Corrosion Science*, 30(4/5): 511-535.

Tunik, L., Addadi, L., Garti, N. and Furedi-Milhofer, H., (1996). Morphological and Phase Changes in Calcium Oxalate Crystals Grown in the Presence of Sodium Diisooctyl Sulfosuccinate. *Journal of Crystal Growth*, 167: 748-755.

Turner, C.W., (1993). Rates of Particle Deposition from Aqueous Suspensions in Turbulent Flow: A Comparison of Theory with Experiment. *Chem. Eng. Sci.*, 48(12): 2189-2195.

Turner, C.W. and Klimas, S.J., (2001). The Effect of Surface Chemistry on Particulate Fouling under Flow-boiling Conditions, *Proceedings of the 4th International Conference on Heat Exchanger Fouling Fundamental Approach & Technical Solutions.* United Eng. Foundations Inc., Davos, Switzerland, pp. 19-26.

Turner, C.W., Klimas, S.J. and Brideau, M.G., (2000). Thermal Resistance of Steam-Generator Tube Deposits under Single-Phase Forced Convection and Flow-Boiling Heat Transfer. *The Canadian Journal of Chemical Engineering*, 78(1): 53-60.

Tyson, W.R., (1975). Surface energies of solid metals. *Canadian Metallurgical Quarterly*, 14(4): 307-314.

Ulmgren, P. and Radestrom, R., (1997). *Calcium Oxalate in Bleach Plant Filtrates*, TAPPI Minimum Effluent Mills, San Francisco, CA, pp. 51-62.

Ulmgren, P. and Radestrom, R., (1999a). An Equilibrium Model of the Solubility of Calcium Oxalate in Sodium Chloride Medium. *Nordic Pulp & Paper Research*, 14(3): 214-220.

Ulmgren, P. and Radestrom, R., (1999b). Solubility of Calcium Oxalate in the Presence of magnesium Ions, and Solubility of Magnesium Oxalate in Sodium Chloride Medium. *Nordic Pulp & Paper Research*, 14(4): 330-335.

Vachon, R.I., Nix, G.H. and Tanger, G.E., (1968). Evaluation of Constants for the Rohsenow Pool-Boiling Correlation. *Journal of Heat Transfer*: 239.

Vadekar, V.V. and Hills, P.D., (2001). Evaporative Heat Transfer to Solutions Containing Dissolved Solids: Effect of Vapour-Liquid Equilibrium and Mass Transfer. *Trans. IChemE.*, 79(A): 477.

Velich, V., Sohnel, O., Costa-Bauza, A. and Kroupa, M., (1992) An isoperibolic reaction twin calorimeter for investigating precipitation. *Crystal Research and Technology*, 27(8): 1133-9.

Vyazovkin, S., (2000). Kinetic Concept of Thermally Stimulated Reaction in solids: A View from a Historical Perspective. *Int. Rev. Phy. Chem.*, 19(1): 45-60.

Walford, S.N. and Walthew, D.C., (1996). Preliminary Model for Oxalate Formation in Evaporator Scale. *Proceedings of South African Sugar Technologist Association*, 70: 231-235.

Walthew, D.C., (1996). Aspects of Evaporator Scale Formation and Control in the South African Sugar Industry. *Proceedings of the Conference on Sugar Processing Research:* 22-43.

Walthew, D.C., Khan, F. and Whitelaw, R., (1998). Some Factors Affecting the Concentration of Silica in Cane Juice Evaporators. *Proceedings of South African Sugar Technologist Association*, 72: 223-227.

Walthew, D.C. and Turner, L.M., (1995). Analysis of Scale from Some South African Sugar Mills. *Proceedings of South African Sugar Technologist Association*: 138-143.

Walthew, D.C., Whitelaw, R. and Mohabir, R., (1997). Chemical Cleaning of Evaporators. *Proc. South Africa Sugar Technol. Ass.*, 71: 199-206.

Warpehoski, M.A., Buscemi, P.J., Osborn, D.C., Finlayson, B. and Goldberg, E.P., (1981). Distribution of Organic Matrix in Calcium Oxalate Renal Calculi. *Calcif. Tissue Int.*, 33: 211-222.

Watkinson, A.P. and Epstein, N., (1970). *Particulate Fouling of Sensible Heat Exchangers, 4th International Heat Transfer Conference*, Versailles, France, pp. HE1.6 1-12.

Watson, L.J., (1987). Heat Transfer Mechanisms in Evaporators. *Proc. Aust. Soc. Sugar Cane Technol.*: 221-227.

Webb, R.L., Eckert, E.R.C. and Goldstein, R.J., (1971). Heat Transfer and Friction in Tubes with Repeated-rib Roughness. *International Journal of Heat and Mass Transfer*, 14: 601-617.

Weissbuch, I., Addadi, L., Lahav, M. and Leiserowitz, L., (1991). Molecular Recognition at Crystal Interfaces. *Science*, 253: 637-645.

Wen, D.S. and Wang, B.X., (2002). Effects of Surface Wettability on Nucleate Pool Boiling Heat Transfer for Surfactant Solutions. *Int. Journal Heat Mass Transfer*, 45: 1739.

Wenzel, U., Hartmuth, B. and Muller-Stainhagen, H., (1994). Heat Transfer to Mixtures of Acetone, Isopropanol and Water under Subcooled Flow Boiling Conditions - I. Experimental Results. *Int. Journal Heat Mass Transfer*, 37(2): 175.

Williams, P.S., Moon, M.H., Xu, Y.H. and Giddings, J.C., (1996). Effect of Viscosity on Retention Time and Hydrodynamic Lift Forces in Sedimentation/Steric Field-flow Fractionation. *Chem. Eng. Sci.*, 51(19): 4477-4488.

Winterton, R.H.S., (1998). Where Did the Dittus and Boelter Equation Come From? *Int. Journal Heat Mass Transfer*, 41(4-5): 809.

Yiantsios, S.G. and Karabelas, A.J., (1998). The Effect of Gravity on the Deposition of Micron-sized Particles on Smooth Surfaces. *International Journal of Multiphase Flow*, 24(2): 283-293.

Yiantsios, S.G. and Karabelas, A.J., (2003). Deposition of Micron-sized Particles on Flat Surfaces: Effects of Hydrodynamic and Physicochemical Conditions on Particle Attachment Efficiency. *Chem. Eng. Sci.*, 58(14): 3105-3113.

Yung, B.P.K., Merry, H. and Bott, T.R., (1989). The Role of Turbulent Bursts in Particle Re-entrainment in Aqueous Systems. *Chem. Eng. Sci.*, 44(4): 873-882.

Nomenclature

A	= Exponential constant, or heating surface area, m^2
a	= Activity or constant in eqn (A6)
a*	= Activity of a saturated solution at a given temperature
B	= Constant related to deposit strength in eqn (6.2), (s^{-1}) or Brix concentration (%)
Bo	= Boiling number ($q / \dot{m} h_{fg}$)
B_M	= Dimensional constant in eqn (A21)
b	= Constant in eqn (6.22)
c	= Solute concentration, mol/L
c*	= Solubility concentration of a sparingly salt, mol/L
c_p	= Heat capacity, J/kg °C
C	= Concentration, %(wt %) (for sugar), mol/L (for COM) or kg/m^3 (for particle)
CF	= Concentration factor (or the degree of concentration)
C_{OH^-}	= Molar concentration of OH^- ions
C_{SF}	= Surface coefficient
Δc	= Concentration driving force (= c – c*), (mol/L)
D,d	= Diameter, m
$\overline{D_L}$	= Log mean diameter ((D_o – D_i)/ln(D_o – D_i)), m
D_m	= Molecular diffusivity, m^2/s
E	= Activation energy, kJ/g mol
F	= Enhancement factor for the Chen model in eqn (A1)
$f_{(2)}$	= Activity coefficient of solute in solution, or the Fanning friction factor
F*	= Activity coefficient of solute at crystal
ΔG	= Overall free energy, kJ/g mol
ΔG_{crit}	= Activation energy for homogeneous nucleation, kJ/g mol
$\Delta G'_{crit}$	= Activation energy for heterogeneous nucleation, kJ/g mol
ΔG_s	= Free energy due to the new surface
ΔG_V	= Free energy due to formation of a new volume
ΔG_v	= ΔG_V per unit volume
g	= Gravitational acceleration, m/s^2
g_o	= Constant in eqn (A21)

ΔH^{Φ}	= Heat of precipitation, kJ/mol
h_{fg}	= Heat of vaporization, J/kg
h'_{fg}	= Modified heat of vaporization, J/kg
h	= Heat transfer coefficient, W/m^2 K
i	= Constants in the Taborek et al. and the Epstein model (eqns (6.8) and (6.12))
J	= Rate of nucleation
Ja	= Jacobs number ($c_p \Delta T_{sc} / h_{fg}$)
K	= Deposition rate coefficient in eqn (6.17)
$K_{a, b}$	= Surface dissociation constants, L/mol
K_c, K_{csp}	= Concentration solubility product, $(mol/L)^2$
K_{con}	= Consolidation rate coefficient, m/s
K_d	= Diffusion mass transfer or particle deposition rate coefficient, m/s
K_G	= Overall crystal growth rate coefficient, m/s
K_o	= Pre-exponential factor in eqn (5.4.2)
K_r	= Surface integration rate coefficient in eqn (6.10), m/s
K_{rem}	= Removal rate coefficient, m/s
K_R	= Deposition rate constant of silica scale in eqn (6.19) (m^2 K/kW·s·$(mol/L)^r$)
K_{sp}	= Thermodynamic equilibrium constant
K_t	= Particle transfer rate coefficient, m/s
K_t^+	= Dimensionless particle transfer coefficient (= K_t/u*)
K_1	= Constant in eqn (6.3)
K_2, K_3	= Constant in eqn (6.4)
K_4	= Constant in eqn (6.5)
K_5	= Constant in eqn (6.6)
K_6	= Constant in eqn (6.7)
K_7	= Constant in eqn (6.8)
K_8	= Constant in eqn (6.9)
K_9	= Constant in eqn (6.11)
K'	= Constant in eqn (6.22)
k	= Thermal conductivity, W/m K, or the Boltzmann constant (=R/N)
k'	= Mass transfer constant in the Epstein model (eqn (6.12))
k"	= Attachment constant in the Epstein model (eqn (6.12)), kg^{n-1} s^2/m^{3n-2}
k_r	= Rate constant, m/s
L	= Heated length of tube, m
M	= Molecular weight
M_R	= Si/Ca molar ratio
m	= Mass deposited on surface, kg/m^2
$\dot{m}$	= Mass flow rate, kg/s
N	= The Avogadro number (6.023×10^{23} mol^{-1}), or constant in eqn (A17)
N_m	= Mass flux normal to the wall
Nu	= Nusselt number (hD / k)
n	= The number of ions in a molecular unit, or the order of crystal growth reaction

P	= Sticking probability in eqn (6.5)
P_d	= Probability function in eqn (6.8)
P_r	= Reduced pressure (absolute pressure/critical pressure), Pa
ΔP_{sat}	= Saturated pressure difference corresponding to ΔT_{sat} in eqn (A4), Pa
Pr	= Prandtl number ($c_p \mu / k$)
p	= Reaction order for silica
q	= Heat transfer rate, W
q_t	= Ionic product of CaOx at time t, $(mol/L)^2$
q_e	= Ionic product of CaOx at equilibrium, $(mol/L)^2$
q"	= Heat flux, W/m^2
R	= The gas constant (= 8.314 J/K mol)
R_b	= Maximum bubble radius, m
Re	= Reynolds number ($\dot{m} D / \mu$)
$R_{f(t)}$	= Instantaneous fouling resistance, m^2 K/kW
R_{TC}	= Total resistance to heat flow based on clean surface condition, m^2 K/kW
R_{TF}	= Total resistance to heat flow based on fouled surface condition, m^2 K/kW
R_W	= Resistance to heat flow of the metal wall, m^2 K/kW
R_f^*	= Asymptotic fouling resistance, m^2 K/kW
r	= Radius, m
r*	= Critical nuclei size, m
S	= Suppression factor in eqn (A1)
Sc	= Schmidt number (= ν/D_m)
SS	= Supersaturation ratio (= c/c*)
S_d	= Particle stopping distance, m
S_d^+	= Dimensionless particle stopping distance (= $S_d u^*/\nu$)
T	= Temperature, °C
ΔT	= Temperature difference, K
ΔT_l	= Temperature difference of fluid along the annulus, K
ΔT_m	= Log mean temperature difference, K
ΔT_{sat}	= Superheat (T_w - T_{sat}), K
ΔT_{sc}	= Subcooling ($T_{sat} - T_b$), K
t	= Time, s
U	= Overall heat transfer coefficient, W/m^2 K
U_b	= Bubble growth rate, m/s
$\bar{u}$,u	= Bulk fluid velocity, m/s
u*	= Wall shear velocity ($= u\sqrt{f/2}$), m/s
V	= Molecular volume, m^3
υ_{fg}	= Vapor-liquid specific volume, m^3/kg
W_t	= Transport flux of particles in eqn 2.32, $kg/m^2 \cdot s$
x	= Quality
z	= Valency of ions

Greek Letters

α_r	= Relative supersaturation ratio (=$\Delta c/c^*$ = SS-1)
β_{CaOx}	= Deposit consolidation factor in eqn (5.4.3) by CaOx precipitation
$\gamma_{(12)}$	= interfacial tension, N/m
Γ	= Constant in eqn (A17/18)
δ	= Wall thickness, m
ρ	= Density, kg/m^3
τ_p^+	= Dimensionless particle relaxation time
τ	= Shear stress, Pa
$\overline{\tau}_s$	= Average fluid shear stress at surface, Pa
σ	= Surface tension, N/m
δ	= Heat transfer wall thickness, m
ζ	= Empirical parameter depending on deposit strength in eqn (6.11)
μ	= Dynamic viscosity, N s/m^2
μ_o	= The standard chemical potential, kg $m^2/mol\ s^2$
μ_1	= The chemical potential of the given substance for supersaturated solution, kg $m^2/mol\ s^2$
μ_2	= The chemical potential of the solid crystal face, kg $m^2/mol\ s^2$
ϕ_c	= Consolidation rate, m^2 K/kW·s
ϕ_d	= Deposition rate, m^2 K/kW·s
ϕ_f	= Net fouling rate, m^2 K/kW·s
ϕ_r	= Removal rate, m^2 K/kW·s
ν	= Kinematic viscosity (μ/ρ), m^2/s
Φ	= Heterogeneous nucleation factor
ψ	= Deposit strength factor in eqn (6.9)
Ω	= Water characterization factor in the Taborek et al model (eqn (6.8))
Ψ_o	= Dimensionless term in eqns (A13-A16) representing value of Ψ at zero subcooling and vapor quality ($\Psi = q''/\Delta T_{sat} h_l$)

Subscripts

a	= Adhesion
b	= Bulk
B	= Boiling
c	= Cold fluids
C	= For clean surface or condition
D	= For dirty (fouled) surface or condition
f	= Fouling, deposit or condensate film
FC	= Forced convection
G	= Gas
h	= Hot fluids

iw,I	= Inside wall
ib	= Value at the incipient boiling point
l	= Liquid
NB	= Nucleate boiling
o	= Outside wall
p	= Particle
pool	= Pool boiling
r	= Removable deposit
s	= Surface
sat	= Saturation
sc	= Subcooling
st	= Steam
t	= Total
tang	= Point of tangency
T	= Total of forced convection and nucleate boiling
v	= Vapor
w	= Wall

APPENDICES

APPENDIX I. SUBCOOLED FLOW BOILING MODELS USED IN THIS WORK

1. Chen Model (Chen, 1966)

$$h_T = h_{FC} + h_{NB} = Fh_l + Sh_{pool} \tag{A1}$$

For subcooled boiling the temperature driving forces for nucleate boiling and forced convection are different and the above equation is replaced by:

$$q / A = [Sh_{pool}(T_w - T_s) + Fh_l(T_w - T_b)] \tag{A2}$$

where F=1 for subcooled boiling (as there is no net vapor generation, i.e., x=0 (Gungor and Winterton, 1986).

h_l is calculated using the Dittus-Boelter equation:

$$\frac{h_l D}{k} = 0.023 \mathrm{Re}_l^{0.8} \mathrm{Pr}_l^{0.4} \tag{A3}$$

h_{pool} is calculated by the Forster-Zuber correlation:

$$h_{pool} = \frac{0.00122 \Delta T_{sat}^{0.24} \Delta p_{sat}^{0.75} c_{p,l}^{0.45} \rho_l^{0.49} k^{0.79}}{\sigma^{0.5} h_{fg}^{0.24} \mu_l^{0.29} \rho_v^{0.24}} \tag{A4}$$

where $\Delta P_{sat} = P_{sat}(T_w) - P_{sat}$

$$S = \frac{1}{1 + 2.53 \times 10^{-6} \mathrm{Re}^{1.17}} \tag{A5}$$

where $\mathrm{Re} = \mathrm{Re}_l F^{1.25} = \mathrm{Re}_l$ (x=0 and F=1).

2. Proposed Model Based on Chen's Approach

The non-boiling coefficient, h_l, is calculated using the modified Dittus-Boelter equation:

$$\frac{h_l D}{k} = a\,\mathrm{Re}_l^{0.8}\,\mathrm{Pr}_l^{0.4} \tag{A6}$$

where a = 0.020 for water and 0.023 for sugar solutions. All fluid properties are determined at the bulk temperature.

The pool boiling coefficient, h_{pool}, is calculated by the Rohsenow correlation:

$$\frac{\Delta T_{sat} c_p}{h_{fg}} = C_{SF}\left[\frac{h_{pool}\Delta T_{sat}}{\mu h_{fg}}\sqrt{\frac{\sigma}{g(\rho_l-\rho_v)}}\right]^{0.33}\left[\frac{c_p\mu}{k}\right]^n \tag{A7}$$

where n=1.0 for water and 1.7 for all other fluids. C_{SF} values were obtained by data fitting.

3. Liu & Winterton Model (Liu and Winterton, 1991)

$$q/A = \sqrt{[Sh_{pool}(T_w - T_s)]^2 + [Fh_l(T_w - T_b)]^2} \tag{A8}$$

where

F=1

$$S = (1 + 0.055F^{0.1}\,\mathrm{Re}_l^{0.16})^{-1} \tag{A9}$$

h_l is calculated using the Dittus-Boelter equation and h_{pool} by the Cooper pool boiling correlation:

$$h_{pool} = 55\left(q''\right)^{0.67} P_r^{0.12}(-\log_{10} P_r)^{-0.55} M^{-0.5} \tag{A10}$$

where M is the molecular weight of fluid.

4. Moles & Shaw Model (Moles and Shaw, 1972)

$$\frac{h_T}{h_l} = 78.5 Bo^{0.67} Ja^{-0.5} \Pr^{0.46} \left(\frac{\rho_v}{\rho_l} \right)^{0.7} \quad \text{(A11)}$$

where Ja is the Jacobs number and h_l is determined by the Colburn equation:

$$\frac{h_l D}{k} = 0.023 \operatorname{Re}_l^{0.8} \Pr_l^{0.33} \quad \text{(A12)}$$

The Prandtl number is calcualted at the mean film temperature. All other properties are calculated at the saturation temperature.

5. Shah Model (Shah, 1983)

$$\text{For } Bo > 0.3 \times 10^{-4},\ \Psi_o = 230 Bo^{0.5} \quad \text{(A13)}$$

$$\text{For } Bo < 0.3 \times 10^{-4},\ \Psi_o = 1 + 46 Bo^{0.5} \quad \text{(A14)}$$

$$\text{If } \frac{\Delta T_{sc}}{\Delta T_{sat}} > 2 \text{ or } \frac{\Delta T_{sc}}{\Delta T_{sat}} \geq 6.3 \times 10^4 Bo^{1.25}$$

then

$$q / A = h_l (T_w - T_b) + h_l (\Psi_o - 1)(T_w - T_s) \quad \text{(A15)}$$

else

$$q / A = h_l \Psi_o (T_w - T_s) \quad \text{(A16)}$$

h_l is calculated by the Dittus-Boelter equation.

6. Bjorge, Hall and Rohsenow Correlation (Bjorge et al., 1982)

The incipient boiling criterion, $\Delta T_{sat,ib}$, under subcooled condition is derived by:

For $r_{tang} > r_{max}$

$$\Delta T_{sat,ib} = \frac{1}{1-N} \left(\frac{1}{4 \Gamma N} - N \Delta T_{sc} \right) \quad \text{(A17)}$$

for $r_{tang} < r_{max}$

$$\Delta T_{sat,ib} = \frac{1}{2\Gamma}\left[1+\left(1+4\Gamma\Delta T_{sc}\right)^{1/2}\right] \tag{A18}$$

where

$$\Gamma = \frac{k_l h_{fg}}{8\sigma T_{sat}\upsilon_{fg} h_{FC}},$$

$$N = \frac{h_{FC} r_{\max}}{k_l}$$

$$r_{\tan g} = \frac{4\sigma T_{sat}\upsilon_{fg}}{h_{fg}\Delta T_{sat,ib}}$$

$r_{max} = 10^{-6}$ (m)

if $\Delta T_{sat} < \Delta T_{sat,ib}$,

$$q / A = q_{FC} = h_l(\Delta T_{sat} + \Delta T_{sc}) \tag{A19}$$

else

$$q / A = \sqrt{q_{FC}^2 + q_{NB}^2\left[1-\left(\frac{\Delta T_{sat,ib}}{\Delta T_{sat}}\right)^3\right]^2} \tag{A20}$$

h_l is obtained by the Colburn correlation and q_{NB} is determined by the Mikic-Rohsenow pool boiling correlation:

$$\frac{q_{NB}}{\mu_l h_{fg}}\left(\frac{g_o\sigma}{g(\rho_l-\rho_v)}\right)^{1/2} = B_M \frac{k_l^{1/2}\rho_l^{17/8}c_{p,l}^{19/8}\rho_v^{1/8}}{\mu_l h_{fg}^{7/8}(\rho_l-\rho_v)^{9/8}\sigma^{5/8}T_{sat}^{1/8}}\Delta T_{sat}^3 \tag{A21}$$

where B_M is a dimensional constant depending on cavity size distribution and fluid properties. For comparison with the model proposed in this work, a single value of 1.89×10^{-14} (for water) was used for all fluids tested.

Appendix II Steam-Side Heat Transfer Coefficient (H_{ST}) and Wall Temperature

The resistances to heat transfer on condensing steam-side are much smaller in comparison to those on the fluid-side (see Fig 5.1.7) and as a result the overall heat transfer coefficients will be mainly controlled by the boiling heat transfer coefficients in the annulus. The steam-side heat transfer coefficient was calculated assuming laminar film condensation (30 > Re) on a vertical plate (Kreith 1973; Incropera & DeWitt 1996):

$$h_{st} = 0.943\left[\frac{g\rho_l(\rho_l - \rho_{st})k_l^3 h'_{fg}}{\mu_l(T_{sat,st} - T_{iw})L}\right]^{1/4} \quad \text{(A22)}$$

where ρ_l, k_l and μ_l are density, thermal conductivity and viscosity of saturated liquid at film temperature, respectively. $T_{sat,st}$ is the condensation temperature of saturated steam at prevailing pressure, T_{iw} is the temperature of tube wall on the steam side and h'_{fg} is a modified latent heat of vaporization.

$$h'_{fg} = h_{fg} + 0.68c_{p,l}(T_{sat,st} - T_{iw}) \quad \text{(A23)}$$

where $c_{p,l}$ is specific heat of saturated liquid at film temperature (T_f).

The condensation rate (condensate mass flow rate) is obtained by:

$$\dot{m} = \frac{q}{h'_{fg}} \quad \text{(A24)}$$

The assumption of laminar film condensation may be checked by calculating Re:

$$\mathrm{Re} = \frac{4\dot{m}}{\mu_l b} \quad \text{(A25)}$$

where b is the tube perimeter (inside).

It is found that most of the data were in the wavy-laminar region ($30 \le \mathrm{Re} \le 1800$), for which Re and h_{st} are re-calculated by:

$$\frac{\mu_l h'_{fg}}{4L(T_{sat,st} - T_{iw})} = \frac{k_l}{(1.08\,\mathrm{Re}^{1.22} - 5.2)(\nu_l^2 / g)^{1/3}} \quad \text{(A26)}$$

$$h_{st} = \frac{\mathrm{Re}}{(1.08\,\mathrm{Re}^{1.22} - 5.2)}\frac{k_l}{\left(\nu_l^2 / g\right)^{1/3}} \quad \text{(A27)}$$

where ν_l is kinematic viscosity of saturated liquid at film temperature.

The average steam-side heat transfer coefficients for all working fluids are given in Table 1A. The small variations in h_{st} observed for each fluid are due to non-uniform steam-side conditions in the experimental run (i.e., variation in the steam pressure due to inherent fluctuations in the steam supply conditions).

Table 1A Average values of steam-side heat transfer coefficients for water and sugar solutions [a]

Sugar conc. (%)	Heat transfer rate (W)	h_{st}, average (W/m^2 K)	SD * (W/m^2 K)	RSD ** (%)	Repeatability (W/m^2 K) ***
0	9078.7	6421.4	61.5	0.96	196.7
25	7784.0	6636.0	24.4	0.37	76.9
33	6858.3	6877.8	112.1	1.63	353.1
40	6163.4	6967.6	61.9	0.89	195.1
50	5506.4	7128.1	113.1	1.59	361.8

[a] heat flux effect data not included

* standard deviation = $\sqrt{\dfrac{\sum (h_{st} - h_{st,avg})^2}{n-1}}$

** relative standard deviation = $SD / h_{st,avg} \cdot 100\%$

*** expected difference between two repeated values ($= t \cdot SD\sqrt{2}$, where t is the student t-value at 95% confidence limit)

The method of estimating the steam-side wall temperature, T_{iw}, is based on a heat transfer analysis of an annular heat exchanger together with a trial-and-error procedure using Excel. Essentially, the method involves first assuming a steam-side wall temperature, then calculating the film condensate temperature T_f by the relationship (Kreith 1973):

$$T_f = T_{iw} + 0.25(T_{sat,st} - T_{iw}) \tag{A28}$$

The physical properties of steam condensate at the film temperature are then determined, which are in turn used to determine h_{st} from eqn A27. Then, the heat flux equivalence concept is used to calculate a "new" wall temperature, i.e.,

$$\frac{\Delta T_m}{1/U_o} = \frac{\Delta T_{st}}{1/h_{st}} = \frac{T_{sat,st} - T_{iw}}{1/h_{st}} \tag{A29}$$

This new wall temperature is then used to repeat the above set of calculations until the assumed wall temperature and the "new" calculated wall temperature converge. Accordingly, the wall temperature on the liquid side can also be calculated based on this procedure. Axial variations of wall temperature along the tube surface are generally ~1.5°C.

Appendix III. Correlations Used to Calculate the Physical Properties of Sugar Solution (Peacock 1995):

1. Density

$$\rho_{sugar} = 1000\left(1 + B\left(\frac{B+200}{54000}\right)\right)\left(1 - 0.036\left(\frac{T-20}{160-T}\right)\right)(kg/m^3) \qquad \text{(A30)}$$

2. Heat capacity

$$c_p = 4.1253 - 0.024804B + 6.7x10^{-5}BT + 1.8691x10^{-3}T - 9.271x10^{-6}T^2(KJ/kg\cdot{}^oC) \qquad \text{(A31)}$$

3. Thermoconductivity

$$k = 1.162222x10^{-3}(486 + 1.55T - 0.005T^2)(1 - 0.0054B)(W\cdot m/K) \qquad \text{(A32)}$$

4. Viscosity (dynamic)

$$\log\mu = 22.46N - 0.114 + \frac{(30-T)(1.1 + 43.1N^{1.25})}{91+T}(cP) \qquad \text{(A33)}$$

5. Surface tension

$$\gamma = 0.07575 - 1.4518x10^{-4}T - 2.3922x10^{-7}T^2 + 1.1x10^{-4}B(N/m) \qquad \text{(A34)}$$

where

B = Brix (%)
T = Temperature (^{o}C)
N = B/(1900-18B)

Appendix IV. Additional Data of Sugar Mill Evaporators (Doherty 2002, Honig 1963)

A. Additional Operating Conditions of Sugar Mill Evaporators

Effect No.	*COM* Conc. (ppm)	SS Level	SiO_2 Conc. (ppm)	*SS* level	*Sugar* Conc. in (%)	Conc. out (%)	Avg Conc. (%)	Pressure (kPa)	Superheat /bpe * (^{o}C)	Residence time (min)
1	40	1.4	250	0.7	20	24	22	225	13	4.4
2	50	2.9	300	1.1	24	28	26	147	14	3.5
3	60	4.6	350	1.5	28	36	32	98	12/1.0	4.1
4	80	10	450	2.4	36	46	41	50	20/1.5	6.7
5	130	18.6	750	8.3	46	70	58	15	26/2.5	14.4

* Boiling point elevation

B. Composition of Cane Juice Solution (Doherty, 2002)

Component	Conc. range (ppm)
Ca^{2+}	11-36
Oxalic acid	25-80
SiO_2	250-750
Aconitic acid	290-1300
Malic acid	460-1120
Succinic acid	60-265
Acetic acid	115-800
Glycolic acid	4-50
Citric acid	290-930
K^+	830-2900
Mg^{2+}	26-260
Na^+	35-100
Cl^-	1000-3500
SO_4^{2-}	100-1025

INDEX

A

B

C

D

E

T

U

V

W

X

Y